T0336208

THE GARDEN POLITIC

AMERICA AND THE LONG 19TH CENTURY
General Editors: David Kazanjian, Elizabeth McHenry, and Priscilla Wald

Black Frankenstein: The Making of an American Metaphor
Elizabeth Young

Neither Fugitive nor Free: Atlantic Slavery, Freedom Suits, and the Legal Culture of Travel
Edlie L. Wong

Shadowing the White Man's Burden: U.S. Imperialism and the Problem of the Color Line
Gretchen Murphy

Bodies of Reform: The Rhetoric of Character in Gilded-Age America
James B. Salazar

Empire's Proxy: American Literature and U.S. Imperialism in the Philippines
Meg Wesling

Sites Unseen: Architecture, Race, and American Literature
William A. Gleason

Racial Innocence: Performing American Childhood from Slavery to Civil Rights
Robin Bernstein

American Arabesque: Arabs and Islam in the Nineteenth Century Imaginary
Jacob Rama Berman

Racial Indigestion: Eating Bodies in the Nineteenth Century
Kyla Wazana Tompkins

Idle Threats: Men and the Limits of Productivity in Nineteenth-Century America
Andrew Lyndon Knighton

Tomorrow's Parties: Sex and the Untimely in Nineteenth-Century America
Peter M. Coviello

Bonds of Citizenship: Law and the Labors of Emancipation
Hoang Gia Phan

The Traumatic Colonel: The Founding Fathers, Slavery, and the Phantasmatic Aaron Burr
Michael J. Drexler and Ed White

Unsettled States: Nineteenth-Century American Literary Studies
Edited by Dana Luciano and Ivy G. Wilson

Sitting in Darkness: Mark Twain's Asia and Comparative Racialization
Hsuan L. Hsu

The Garden Politic

Global Plants and Botanical Nationalism
in Nineteenth-Century America

Mary Kuhn

NEW YORK UNIVERSITY PRESS
New York

NEW YORK UNIVERSITY PRESS
New York
www.nyupress.org

© 2023 by New York University

References to Internet websites (URLs) were accurate at the time of writing. Neither the author nor New York University Press is responsible for URLs that may have expired or changed since the manuscript was prepared.

Please contact the Library of Congress for Cataloging-in-Publication data.
ISBN: 9781479820122 (hardback)
ISBN: 9781479820153 (paperback)
ISBN: 9781479820184 (library ebook)
ISBN: 9781479820160 (consumer ebook)

New York University Press books are printed on acid-free paper, and their binding materials are chosen for strength and durability. We strive to use environmentally responsible suppliers and materials to the greatest extent possible in publishing our books.

Manufactured in the United States of America

10 9 8 7 6 5 4 3 2 1

Also available as an ebook

For David, and for Marguerite,
who likes to say hello to flowers

CONTENTS

LIST OF FIGURES

Introduction

A Case for Plants

"Not a Sedentary Study"

In 1829 a teacher named Almira Hart Lincoln Phelps published a textbook that changed the course of botanical instruction in the United States.[1] By encouraging students to seek out plants they could find growing in nearby meadows, forests, and yards, *Familiar Lectures on Botany* made systematic botany accessible to a young, and notably female, readership. Her pupils were among the first generation of girls in the republic to receive a formal botanical education, and Phelps wrote *Familiar Lectures on Botany* in a manner that would prove appealing to beginners; over time, the book went through twenty-eight editions and sold roughly 350,000 copies.[2] Reviewers praised the book as a credit to the nation for the way it guided students toward an appreciation of order and moral feeling.[3] Its lessons might easily translate to the management of the home, seen as the particular purview and responsibility of women, and thereby help strengthen the nation's development. The book's broad appeal reflected the growing scientific literacy and curiosity of middle-class Americans and heralded the beginnings of a monumental craze for horticulture that would sweep the antebellum United States.

While rooted by local botanical observations, Phelps's opening image for this germinal volume nonetheless invited her readers to think about plants in a global context. Across from the title page of *Familiar Lectures*, Phelps placed an engraved illustration depicting Chimborazo, an Andean peak towering more than twenty thousand feet above sea level, alongside its actively volcanic neighbor Cotopaxi. Through a combination of text and image, this scientific engraving presents the distribution of plants around the globe via an elevation of these two

Figure I.1. "Progress of Vegetation at Different Elevations" placed across from the title page of *Familiar Lectures on Botany* (1829). Image taken from a second edition of the book (1831) in the Biodiversity Heritage Library. Contributed by the Pennsylvania Horticultural Society.

mountains. The slopes of the mountains act as a canvas for the compact illustration of global climatic regions, starting with the tropics at the base of the mountains and culminating with polar ice at their peaks. In between, the image depicts zones in an ascending progression featuring the names of relevant plant species: the lower slopes hold tropical plants, a little higher up are plants of a temperate climate, and at the top of the mountain there is only "Snow" and "Perpetual Snow." In this way, Phelps shows an extended botanical lineage, linking the local and the familiar to the far-flung and exotic, effectively collapsing a multiplicity of ecosystems into a chart comprehensible to fledgling American botanists.

From the beginning, *Familiar Lectures on Botany* emphasizes a global scientific perspective, choosing to showcase plants found around the world. At the same time, the book makes a specific case for botany's relevance to the young American women for whom it was written. Praising botany as a science that is "peculiarly adapted to females" because "the objects of its investigation are beautiful and delicate," Phelps delights in describing its less delicate collection practices. She writes in her introduction that the discipline "is not a sedentary study which can be acquired in the library, but the objects of the science are scattered over the surface of the earth, along the banks of the winding brooks, on the borders of precipices, the sides of mountains, and the depths of the forest."[4] To study plants, she underscores, is to explore both one's own backyard and also the flora to be found on the banks, borders, and mountainsides of the globe—in short, to approach even the most familiar of terrain as situated within a global network.

Curiously, Phelps never mentions the striking diagram of Chimborazo and Cotopaxi in her text. Perhaps this was because by 1829 she had no need of doing so: many of her readers would have recognized it as a near copy of a famous illustration by Prussian naturalist Alexander von Humboldt. Already a celebrated figure in Europe and America for his best-selling books about his travels throughout Latin America, Humboldt was one of the preeminent explorers of the early nineteenth century, and many of Phelps's well-heeled readers would have had a copy of his writings on their family's bookshelves. As the mountain diagram would suggest, one of Humboldt's most important insights was his recognition that similar plants grew at similar latitudes around the planet. Humboldt called the image he created to represent this observation his "Naturgemalde" or "painting of nature."[5]

Such insights depended on the rapidly developing infrastructure of empire, which facilitated the voyages of naturalists like Humboldt. Although Humboldt himself was a critic of colonialism who paid his own way to the Americas, his scientific exploration of the Spanish colonies required Spain's approval and took place within the broader context of imperial missions to assess the economic value of the New World's flora and fauna.[6] Bioprospecting expeditions were a vital imperial tool in the eighteenth and nineteenth centuries, allowing colonial powers in Europe to catalogue potentially useful plants around the globe. Humboldt's original Naturgemalde listed both species that were economically useful and those that were not, but the version Phelps included in her book focuses on plants of commercial and imperial interest. In the tropical zone of the mountain as it appears in *Familiar Lectures* are palms, spices, sugar cane, coffee, tea, and indigo—all plants of foundational importance to the wealth building of nineteenth-century empires—along with the breadfruit trees that fed enslaved plantation workers. Unsurprisingly, these crops were typically excluded from the polite botany encouraged by moralists as appropriate for women's study since they were tainted by the violent racial oppression on which empires were built.

For nineteenth-century Americans, crops such as these represented the most salient political context for plants. Europeans and colonial Americans had long framed plantation cultivation as a project that managed and therefore tamed and civilized inferior bodies and landscapes. As postcolonial scholars have illustrated, imperial discourse centered white European and America men as standard-bearers of culture, knowledge, and power in this global network of people and places.[7] Yet the heat and humidity of the tropical climate of the plantation zone presented an acute challenge to European ideas of bodily integrity.[8] Planters relied upon controlling nature—tilling soil, weeding fields, delineating borders—not only to yield crops but also to manage perceived threats to their persons. Enslaved people, Maroon communities, and Indigenous groups used knowledge of local environmental conditions to resist and disrupt the institutionalized violence of plantations.[9]

New England homes may have been set far away from tropical plantations, but they were governed by many of the same imperatives. Women in charge of domestic households were expected to exert a civilizing influence on their families and to raise children who would be good citizens. As a set of didactic ideologies, domesticity helped produce the raced and gendered idea of the discrete, sympathetic body.[10] One of the reasons botany rose in popularity so quickly is that its classification practices reinforced the idea of the discerning subject appreciating natural objects, shoring up ideas of Enlightenment order. It also emphasized that home was a place for managing plants and animals alongside people. A close ally of the ideology of domesticity, in more than just etymology, was the set of practices known as domestication. Bringing land under cultivation or selectively breeding fruits and flowers for desired traits promised more orderly, productive, and pleasing environments.[11]

The sketch of Chimborazo illustrates the proximity of these projects, placing plants like ivy and nettle, familiar to the middle-class New England homemaker, in relation to crops like indigo and cotton. And while few of Phelps's readers would ever have been able to make a trip to the Andes, by the time she published *Familiar Lectures*, the proliferation of imperial bioprospecting networks meant that a gardener in rural New

England could order plants for the home from almost any continent. Seed catalogues compressed the world's flora between their covers, and as a result foreign flowers increasingly populated her readers' gardens: middle-class Americans could enjoy the fruits of decades of imperial botany. Humboldt, Phelps, and their readers lived in a world where botany served competing empires, but the picture of Chimborazo renders the world's plants without geographic or political borders, making the side of a mountain look like a giant garden plot.

The unexpected appearance of an Andean mountain at the start of a beginner's botany book speaks to the argument at the core of *The Garden Politic*: American gardeners understood the plants they encountered at home as political objects swept up in the scientific and environmental changes of nineteenth-century imperialism.[12] One key point here is that these Americans engaged with plants *at home*. Millions in the early republic interacted with their environments by planting with their own hands in gardens and greenhouses.[13] Political environmentalism in nineteenth-century American culture is usually associated with wild or sublime landscapes set against the trappings of the domestic and the everyday.[14] While ecocritics have productively recovered the political importance of spaces like swamps that are resistant to human control, the home has often appeared as a kind of orderly foil to the prodigiousness of uncultivated spaces. This book shows that, as tidy as domestic environments might appear in theory, actual attempts to domesticate plants made farmers and gardeners acutely aware of the unpredictable agency of plant life.[15] The seeds they planted came from all over the world, and as Americans acknowledged this botanical diaspora, different gardeners came to different political conclusions. For some, the movement of plants made settler colonialism itself seem natural.[16] At the same time, however, other Americans saw new meanings in traveling plants, meanings that could challenge existing structures of power. Botanical instruction like that in Phelps's book was meant to bolster existing social and political hierarchies, but new scientific ideas combined with the experience of caring for their own plants encouraged American gardeners to consider other ways of organizing society. Plant agency had social and political consequences because, as Philip Pauly has shown, the American understanding of culture was biotic, an understanding that itself shaped ideas of societal improvement.[17] Plants did not always obey human plans, troubling the grand visions of control held by those in political power and the hopes of home gardeners in their own yards. If domesticators sought to fully control plant life, plants regularly disrupted these expectations and frustrated human desires.[18] The liveliness of plants, their tenacity as weeds and fickleness as valuable cultivars, and the ways in which they challenged human efforts to control them influenced how a generation of gardeners reflected on the meanings of race, empire, and home.

Each chapter of *The Garden Politic* takes up a different facet of domestic plant cultivation. In doing so, the chapters show how nineteenth-century Americans engaged with plants to understand present political realities and to imagine alternative political futures. Each chapter is anchored around amateur gardeners who were also significant authors of the mid-nineteenth century, including Lydia Maria Child, Nathaniel Hawthorne, Emily Dickinson, Harriet Beecher Stowe, and Frederick Douglass. These authors, alongside others including naturalists and periodical writers, discerned the importance of horticultural practices to issues of national economy and governance. Their own horticultural knowledge made them astute commentators on changes in plant culture across the century. As observers and practitioners, they reflected the experience of being enmeshed in a global botanical network and provide a window into how shifting horticultural ideas shaped the terms through which nineteenth-century Americans made and legitimated their political realities. These writers also used their literature to theorize how the intractability of plant life could disrupt some of the most potent biopolitical aspects of cultivation. Living in a political system that exploited people and material ecosystems, they recognized alternative models of relating to other living beings.[19]

While the book focuses on political issues in the context of the nineteenth-century United States, it illustrates the global genealogies of the organisms with which Americans came into contact every day. I use "global" here to describe connections that Americans felt and imagined to wide-ranging latitudes and distant continents, rather than their sense of a totalizing whole.[20] That is not to say that naturalists and authors never entertained holistic ideas, as Humboldt's sketch of Chimborazo attests.[21] But they were as interested in the material compression of time and distance, which could be felt by growing a multiplicity of plants "originally" from other continents in their own yards and greenhouses, as they were in theorizing the globe as a whole.[22] Thus Dickinson imagined a floral landscape extending from Amherst to Kashmir, and Douglass described how new horticultural technology allowed the planter class to consolidate the harvest bounty of a range of climates on a single plantation.

In what follows I trace the nineteenth-century meanings of plants at three key interwoven scales: the global scale of imperial bioprospecting, the national scale in which plants were essential to establishing American political power and identity, and the local scale of the domestic garden. Americans of the nineteenth century understood plants as objects that worked at all three scales at once. Humboldt's mountain loomed over the activities of everyday gardeners, shaping the home as a key site of environmental thought and practice. But despite the fact that the domestic is "forged in the crucible of foreign relations," to quote Amy Kaplan, ecocriticism has still largely considered domestic environments to be immune to what Kaplan calls the "anarchy of empire."[23] As we will see, gardeners routinely perceived and made meaning out of this anarchy.

Plants and Empire

In 1829, the same year Phelps published her best seller, the English doctor Nathaniel Bagshaw Ward invented an effective means for protecting plants from the "fuliginous," or sooty, London air. In doing so he inadvertently created an essential tool in the development and expansion of European empires.[24] Experimenting at home with a moth cocoon in a sealed glass jar, Ward discovered that a fern sprouting in the glass was able to thrive. Ward used this knowledge to develop a glass case for growing plants that might protect them from the elements. Within a few decades, Wardian cases were adopted for the transport of plants on long ocean journeys. Set on the decks of ships, these cases allowed delicate plants to get the sun they needed while remaining protected by glass from exposure to salt water, fluctuating temperatures, and the harsh weather of transoceanic passages. While most plants had previously succumbed to the inhospitable conditions on ships, Ward's case allowed for a far greater number of plants to arrive at their destination alive, boosting the commercial, scientific, and political circulation of plant matter around the globe.

Figure I.2. "On the Conveyance of Plants and Seeds on Ship-Board" from Ward's *On the Growth of Plants in Closely Glazed Cases*. Image taken from the Biodiversity Heritage Library. Contributed by New York Botanical Garden, LuEsther T. Mertz Library.

One famous example of this circulation was the cinchona tree, the bark of which could be boiled down into the powerful antimalarial quinine. Endemic to the eastern slopes of the Andes, cinchona harbored antimalarial properties that were first discovered by Indigenous groups in the region. When Spain invaded in the sixteenth century, it sought control over this wonder drug, and the Crown later established a royal reserve. Realizing the plant's medical significance, the French, Dutch, and British smuggled specimens out of this Spanish-controlled region and sent them to government-sponsored gardens and plantations. The British sent theirs to India, seeking to produce a ready store of quinine that further facilitated imperial efforts to colonize other areas of the globe where malaria was endemic.

Cinchona was one of many strategic plants whose distribution and control were an obsession of European empires. Tea, sugar, tobacco, and opium were the basis for colonial economies organized around plantations.[25] Empires also worried about many less celebrated plants, like the breadfruit moved from the East to the West Indies as a potential cheap source of calories for enslaved workers.[26] These plants were recognized as the material resources upon which political and economic power could be erected, and their cultivation had to be protected at all costs. Full-grown plants, tender slips, and seeds all held political value. Plantations emerged as the basis for imperial power during the seventeenth and eighteenth centuries, organized around the violent subjugation of life—human, animal, and plant. Scientific classification projects reinforced the hierarchies of plantation life, and enslaved laborers were racialized, a biopolitical process of subordination.

The rhetoric of cultivation justified and helped conceal these processes: by cultivating land colonists believed they were improving it and managing the threats inherent in the natural world. Eighteenth- and nineteenth-century artists frequently depicted colonial plantations as idyllic gardens in which the unruliness of nature had been tamed.[27] Scientific representations could likewise recast plantation realities by lifting plants out of their local contexts. Unrepresented in Humboldt's image of Chimborazo and Cotopaxi were the Indigenous polities dispossessed through colonial settlement. Nor does it depict the human labor of subordinated populations required to make plantations productive.[28] Botanical knowledge could conceal as much as it revealed.

The grandest symbol of this process of elision in the British Empire was the Royal Botanical Gardens at Kew outside London. Kew fulfilled a critical function in maintaining and expanding the empire's plantations. Plants from as far away as Australia were sent to Kew, where their economic potential was ascertained, and valuable specimens were sent to suitable climates within the empire for cultivation. Kew functioned not only as a laboratory for prized plants but as a public symbol of the empire's control over nature. As with the Wardian case, developments in greenhouse technology made it increasingly possible to artificially regulate climate. In his book *On the Growth of Plants in Closely Glazed Cases* (1842), Ward had offered his "firm conviction that, with the progress of science, any climate on the face of the earth will be readily imitated and maintained."[29] At Kew, the "palm stove," a greenhouse for tropical plants, was the heart of this enterprise. When it was finally finished in

1848, it was hailed as an iron and glass symbol of Britain's engineering prowess and admired as an organic utopia designed to educate and delight. Visitors could climb spiral staircases in the center dome and gaze down upon living specimens of tropical plants collected across the vast and rapidly growing British Empire. Below them, palm trees, bamboo, a variety of grasses, and other tropical plants formed a lush canopy, offering an impressive showcase of the world's greenery and the botanical possibilities of imperial bioprospecting. While the primary intention of the display was aesthetic grandeur and scientific education, the chosen plants also advocated for the usefulness of botanical collection to British interests: sugar cane, for instance, made an appearance in the greenhouse, along with banana and plantain trees and other plants whose edibility immediately demonstrated the value of their acquisition. Like Humboldt's Naturgemalde, the palm stove gave a singular impressive picture of plants grown around the globe.

In the nineteenth century, the railroad, telegraph, steam engine, and photography were commonly credited with the ability to annihilate space and time.[30] But so could the garden.[31] Greenhouses, Wardian cases, and even seed catalogues that advertised specimens from across the globe inspired a sense that global distances were being radically compressed and controlled. By the middle of the nineteenth century, a gardener sitting at home in Britain or the United States could place a single order for seeds that originated in South America, Asia, Europe, and Australia. If it was not quite possible to cultivate pineapples in Uppsala, as the Swedish botanist Carl Linnaeus had hoped in the previous century, it was possible by the mid-nineteenth century to grow tropical plants under glass in a heated New England home. Describing the plants in her greenhouse, Emily Dickinson wrote to dear friends, "My flowers are near and foreign, and I have but to cross the floor to stand in the Spice Isles."[32] When gardeners and writers confronted novel biota, the experience inspired them to imagine a sense of intimacy with distant parts of the globe.

Plants and Nation

As the global traffic of plants brought Westerners into biotic contact with distant places, naturalists and writers in Britain and America saw the diversity of colonial flora and fauna as a way to assert regional and national distinctiveness.[33] Assessing the biotic potential of North America, eighteenth-century naturalists on both sides of the Atlantic spoke of the continent's unique character. Early American botanists like John Bartram and his son William were embedded in an extended network of European naturalists interested in the potential value of plants growing across North America. As naturalists and bioprospectors in the late eighteenth century and early nineteenth had enumerated the distinctive biota to be found in North America, they helped foster a now-familiar founding narrative about how the national soil and climate might yield an equally exceptional body politic.

Politicians and writers distinguished American biota in order to conceptualize the United States as an exceptional political territory.[34] And while many naturalists praised the new species they found growing on the North American continent, others described a degraded American nature. The French naturalist George-Louis Leclerc, Comte de Buffon took a dim view of the New World, believing that the climate caused life to degenerate. Responding to Buffon, Thomas Jefferson offered a meticulous catalogue of American flora and fauna and compared them advantageously to those found in Europe, a claim that he extended to "the man of America, whether aboriginal or transplanted."[35] At the same time, though, he worried that he could not successfully grow wine grapes on par with those in Europe and wondered what that might signal for the cultural vitality and longevity of the republic.[36] The late eighteenth-century writer Hector St. John de Crevecoeur in *Letters from an American Farmer* (1782) memorably opined, "Men are like plants; the goodness and flavor of the fruit proceeds from the peculiar soil and exposition in which they grow." Soil, in other words, is the basis for character. Decades later Henry David Thoreau too espoused America's biodiversity as uniquely advantageous for the republic. In his late essay "Walking" Thoreau asks, "Where on the globe can there be found an area of equal extent with that occupied by the bulk of our States, so fertile and so rich and varied in its productions, and at the same time so habitable by the European as this?" As an authority on the vigor and variety of American natural "productions," he cited Humboldt.[37]

White settlers had long used the Lockean idea of improving the land through cultivation to justify colonization along the Atlantic coast of

North America. Like descriptions of exceptional flora and fauna, in the late eighteenth and early nineteenth centuries such settler cultivation projects aligned plants with the republican mythos.[38] On the Eastern Seaboard, the founding fathers sought to illustrate the potential of their democratic experiment (and claim to the land) through experiments with cultivation.[39] They believed the body politic depended on agriculture and could be understood in agricultural terms. Jefferson imagined a democracy of yeoman farmers, while he and other wealthy planters relied on slavery to facilitate and sustain plantation agriculture. Thriving plantations confirmed their sense of the vigor of the young nation and justified the enslaved labor used to make the fields and gardens yield saleable abundance. Such a vision made national identity inextricable from plantation slavery.[40] It also hid the ways that white colonists depended upon the botanical practices and expertise of enslaved workers, Indigenous communities, and free people of color.[41] When planters and bioprospectors identified valuable plants, they frequently sought to translate them into the universalizing language of binomial nomenclature, which allowed them to more easily control their meanings within national, transnational, and imperial frameworks.[42]

Both wild and cultivated plants, then, sustained ideas about the republic's peculiar character, forming the basis for botanical constructions of nationalism. As biotic exchange accelerated, naturalists and amateur collectors understood that acquiring plants from around the globe was a national project just as it had been an imperial project in Britain. While the United States never developed a national garden on the scale of Kew's, gardeners appreciated the relationship between plant acquisition and power. The influential and long-running *American Agriculturalist* praised Kew's promotion of science, art, medicine, commerce, agriculture, horticulture, and manufacture, noting, "The government employs plant collectors in every part of the world [. . . to create] a huge encyclopedia, printed with facts instead of words!"[43] Such adulation speaks of American aspirations toward a similar kind of comprehensive control, and over the course of the nineteenth century the U.S. government sought to strengthen its position through the collection and commodification of plant knowledge. By the late 1830s the U.S. Patent Office had taken on the task of collecting seeds and managing agricultural improvement. The Patent Office building contained a greenhouse and garden dedicated to plant research and saw seed distribution to farmers as one of its most important tasks.[44] This federal investment in cultivation mirrored the rise of botanical instruction in schools and home gardening among the middle class. As we will see, the political economy and social project of cultivation often went hand in hand.

Botany and Domestic Order

The imperial and national contexts for botany facilitated the rise in domestic gardening. While the story of nineteenth-century domesticity has been told many times, we have yet to appreciate how plant culture unsettled some of the most powerful domestic ideologies of the antebellum period. The disruptive power of plants in this regard and the challenges they present to today's prevailing critical narratives can be understood through the way botany's taxonomic project broadly upheld conservative ideas. As a set of ideologies and cultural practices that imposed a sense of order on the world, domesticity could oppose an intimate private sphere against a public one, delineate cultivated space against wild growth, or signal a separation between the familiar and the foreign.[45] Botany's taxonomic conventions helped neatly organize the world's flora, but more broadly botany taught a habit of mind that elevated the idea of orderliness and system. In the opening chapter of her *Familiar Lectures*, Phelps insists on botany's practicality and the applicability of its lessons to the "most common concerns and operations of ordinary life."[46]

Botany's ability to inculcate order derived from its methods of classification. In the mid-eighteenth century, Swedish naturalist Carl Linnaeus developed a taxonomic system for classifying plants based on their sexual organs. The simplicity of this system made it attractive as an instructional tool, and Linnaean taxonomy became popular in Britain during the late eighteenth century as a way to characterize sexual difference. The physician and naturalist Erasmus Darwin, for instance, grandfather to Charles, wrote a long poem to teach Linnaeus's system through sexualized courtship language. "The Loves of the Plants" (1789) describes the pistils of plants as demure young ladies and the stamens their male suitors. From Ovid to Shakespeare, writers had long compared humans to flowers, but Linnaeus's system grounded these social observations in the relatively new authority of science. Victorian writers likewise turned to organic metaphors to naturalize courtship. "Blooming" young women appeared in novels from Austen to James, with their fertility—and thus their marriageability—signaled by their alignment with such a natural botanical process.[47] At the same time, specific flowers were readily associated with particular sentiments. "Language of flowers" books created a symbolic system for understanding the emotional meaning of particular plants, fostering floral metaphor's sentimental value. Emotion could be classified, mapped, and expressed through floral exchange.

Almira Phelps sought to be accessible when she wrote *Familiar Lectures*, and she adopted Linnaean classification because it was easy to learn and use, helping students see larger patterns. As the first science fully integrated into school curricula for women and men alike in the new republic, botany taught students to observe, collect, and classify the natural world around them and to derive larger lessons about God's design from these observations.[48] In her introduction to the book Phelps extolled how "the Deity has . . . given to our minds the power of reducing [objects] into classes, so as to form beautiful and regular systems, by which we can comprehend, under a few terms, this vast number of individual things."[49] Phelps joined many moralists, philosophers, scientists, and artists on both sides of the Atlantic who praised the close study of plants because it was considered to be a pious activity that could illustrate the relationship among all of God's creations in an ordered universe.[50] For instance, naturalist William Bartram understood his botanical observations in terms that were both religious and domestic. Early in his 1791 *Travels*, recounting years of traveling and botanizing in the southeastern United States, he characterizes the prodigious variety of the natural world as fitting within "a glorious apartment of the boundless palace of the sovereign Creator, . . . furnished with an infinite variety of animated scenes."[51] Bartram's architectural conceit renders the whole of the globe a kind of domestic interior decorated with useful and pleasing plants, not unlike the Wardian case.

To appreciate God's work was to open up all of one's senses for instruction—seeing, touching, and smelling—that would both "aid the *memory* and *understanding*" and pique "*emotions*, too."[52] Phelps encouraged students to couple sensory observation and emotion, and to marry an aesthetic and pragmatic understanding of plants. Her subsequent

Botany for Beginners (1833) exhorted students to think of "his love and kindness who causeth the earth to bring forth, not only 'grass for the beast of the field, and food for the use of man,' but a rich succession of curious and lovely blossoms for our admiration and enjoyment."[53] In short, God's order was not simply natural but also beautiful, meant to be enjoyed by those with appropriately trained senses.

In encouraging her readers to find the value of botany in its ability to help them categorize and generalize, Phelps stressed that her lessons had social value. Society, rooted in God's design like the natural world, was ordered, and botany had "without a doubt, a tendency to induce in the mind the habit and love of order."[54] Both scientific and sentimental floral classification encouraged the idea that there was a "natural" order of society. The job of the domestic woman was to help cultivate it.

The metaphor of cultivation functioned as one of the most powerful ways to socialize young men and women into this order. Moralists encouraged readers to cultivate such various traits as good habits, strong morals, religious sentiment, temperate feeling, a fine taste, and the right social connections. Books extolling the cultivation of the self were a cottage industry in the nineteenth century, and the practice of self-cultivation was frequently imagined through botanical tropes, as in Louisa May Alcott's *Flower Fables* (1854–1855), which presents a series of moral lessons about selflessness, humility, and gratitude through the fate of flowers in a fairy land. Analogies between gardening and child-rearing were plentiful in popular periodicals from the 1830s and 1840s. One such magazine offered children the choice between being "like a bad and *poisonous plant*, which everybody avoids and wishes away, or a good and wholesome plant, which we cultivate carefully, and love to have near us."[55] Another cautioned its young readers that without careful cultivation, they could expect only "the most luxuriant growth of unruly appetites," including "*avarice*, like some choking weed," and "*revenge*, like some poisonous plant, replete with baneful juices."[56] Readers who grew up on these comparisons knew what happened to an untended garden.

In fact, one of the best-selling books in the early nineteenth-century United States was a French novel about a political prisoner who falls in love with a plant and thereby learns the moral virtues of cultivation. In X. B. Saintine's *Picciola* (translated into English in 1838) a French nobleman is jailed for an assassination attempt on Napoleon and redeemed only through his relationship with a flower that grows through the stones of his prison yard. Through observing and caring for this flower, he develops just the empirical and sensitive abilities botany was supposed to instill. American periodicals praised the novel for its excellent "moral bearing" and for "assail[ing] the secret infidelity which is the bane of modern society."[57] The novel remained popular in the United States for decades and was included alongside works like *Robinson Crusoe*, *Gulliver's Travels*, and *The Vicar of Wakefield* in an 1873 anthology of famous fiction to which Harriet Beecher Stowe wrote the introduction.[58] Its enduring status speaks to the power of plants as object lessons for appropriate behavior.

Yet for all that cultural authorities turned to plants to discipline, they just as often experienced plants as intractable. Saintine's neat narrative sustains the flowering plant and its lesson for his protagonist, but many gardeners' experiences with plants countered or complicated the easy lesson. In the context of global circulation, plants blurred boundaries. Gardening and horticulture often unsettled the distinction between domestic and foreign, as plants from around the world increasingly entered domestic spaces or became domesticated. Gardeners turned to seed catalogues as the source of novel new plant varieties. Many of these catalogues noted the provenance of plants in their collection, listing countries and regions all over the globe. Some even detailed specific mountains or lakesides where a plant was first "discovered." The portability of plant specimens illustrated how much the environment was in flux and how unnatural were the borders—political and social— that regulated their circulation.[59] The movement of plants also collapsed the idea that domestic and wild were completely oppositional. While the distinction of "domestic" versus "wild" shaped the development of modern classification systems, home gardeners routinely blurred these distinctions, bringing "wild" plants into cultivated beds and spreading domesticated seeds into uncultivated landscapes.[60]

Some plants made themselves too at home after purposeful introductions or were introduced accidentally as a by-product of commercial activity. For all the energy devoted to maximizing the presence of unique and desirable plants within the ever-expanding political borders of the nineteenth-century United States, politicians, bureaucrats, scientists, and home gardeners were constantly surprised by unintentional biotic introductions and unpredictable plant behavior. Jimsonweed, for instance, robustly self-propagated along the Massachusetts shoreline as an accidental by-product of commercial trade. Thoreau reveled in its introduction, calling it a "cosmopolitan weed" that "secretes itself in the holds of vessels and migrates." Thoreau's depiction of this "cosmopolite and veteran traveler," a "Captain Cook among plants," makes cosmo-politanism a botanical quality, one that challenges human control and even human historical accounts. "What historian knows when it first came into a country!" he archly exclaims.[61] Thoreau's depiction delights in the difficulty plants posed to official narratives. No matter how inten-sively politicians and horticulturalists managed the circulation of plants, the plants themselves behaved unpredictably: they traveled beyond the sites of their initial plantings, took root in unexpected places, failed to thrive, or grew too well and crowded out other more desirable species. The adaptability and tenacity of such plant life in turn fascinated and frustrated gardeners, who understood through daily experience how plants could undermine human control, from migrating beyond prop-erty boundaries to spreading unbidden across political borders.

Plants that moved of their own accord vexed some and filled others with awe, and these reactions were routinely recorded in the pages of popular periodicals. One 1854 *Putnam's* article marveled at plants' ex-traordinary ability to travel—through air, water, and earth. Everywhere a person might look, "he sees plants quietly and mysteriously perform their humble duty in the great household of nature. Plants alone—it would at first sight appear—have no home, for they seem to be at home everywhere."[62] In this way plants actually mimicked people on the move: at the start of "Walking" Thoreau contemplates the meaning of the word "saunterer," "from *sans terre*, without land or a home, which, therefore in the good sense, will mean, having no particular home, but equally at home everywhere."[63] This idea of being "at home everywhere" upends the idea of a bounded arena for thought or practice, opening up the idea of cosmopolitan connections with faraway places.[64] Plants could inspire a sense of adaptability and expansiveness through their travels. Seen in this light the home is mobile and protean, not aligned with any bounded sphere of human activity—it is a liberating idea, and a colonizing one.

Beyond the ways that plants flouted boundaries, gardeners drew other surprising lessons from their behaviors. With life cycles that sometimes lasted only a few days and other times extended far beyond human ones, plants modeled life differently. Many who cared for plants at home expressed a sense of kinship with them, prompting questions about how to characterize this other form of life. Scientists across the nineteenth century debated the extent to which plants could feel and speculated over the possibility of their sentience. These debates found a receptive audience among periodical readers who were keen gardeners and horticulturalists. Those who spent their days tending to plants were unsurprised by claims that sunflowers kept time by following the sun or that mimosas exposed to acid experienced pain.[65] They had seen sunflowers turn on their stalks and witnessed mimosa leaves shrink from touch, for the home garden provided a test plot for careful observation. Even Charles Darwin, the most famous biological voyager of the nineteenth century, spent a significant portion of his life experimenting with plants and animals in his own home.[66] Domestic experiments instructed home gardeners in the ways that plants could be unpredictable and disruptive, making them agents rather than passive receptacles of human desire.

In other words, the lively materiality of plants mattered to the people who engaged with them. A number of scholars have thoroughly demonstrated how nineteenth-century scientists, philosophers, and literary authors turned to the natural world to justify exclusionary practices or to form the basis of social critique.[67] This book makes a different, if related, claim. In focusing specifically on plants, I recover historically specific debates that operated not at a register of this more capacious category of "nature," the "natural world," or the "environment" but at the more intimate level in which authors and their audiences categorized plants as a unique form of life. My exclusive focus on plant life reveals the close and transformative attachments that gardeners had for plants in their care. The intimate scale at which plants became "part of the family," as Hawthorne once mused in his notebook, meant they were not always simply synonymous with an abstract nature.[68]

This intimate scale also helps us see how a universalizing logic still operates in much contemporary discourse about the environment.[69] The recent "plant turn" in the environmental humanities has called attention to Western societies' consequential treatment of plants as useful symbols and valuable commodities but relatively inert matter. While tropes like rootedness have long organized social and political belonging, such metaphorical use comes at the expense of considering plant forms in their own right.[70] Michael Marder has revealed the extent to which continental philosophy has made use of plants as conceptual projections that flatten their liveliness.[71] Jeffrey Nealon has likewise shown how Foucault and later theorists of biopower forget plant life. He illustrates how even those who analyze political control over life have repeatedly defined "life" narrowly. The field of animal studies has attempted to correct this oversight, and scholars in the newer field of plant studies are working to highlight the ways in which plants become co-opted to human agendas.[72] A number of recent works focus seriously on plant life, seeking to create space for their otherness. Natania Meeker and Antónia Szabari have used the term "radical botany" to characterize the destabilizing and speculative role of plants in Western writing dating back to the seventeenth century. Along similar lines critics have readily described how writers have invoked plant life, from the scientific to the science fictional, to queer or disrupt normative values.[73]

One of the key challenges for scholars working on plant studies is the extent to which it is possible to represent this dramatically different form of life without slipping into anthropomorphism. But anthropomorphism is not a monolithic category, and postcolonial and critical race theorists have cautioned wariness of the categorical "anthro." As we will see, nineteenth-century Americans sought to make meaning out of both the similarities and the differences that plants represented. In recovering the complex iterations of nineteenth-century encounters between plants and people, I consider how a fascination with plant life impacted political beliefs. Plants (and animals) were inextricably bound up in political worldmaking, often through their treatment as passive objects. The experience of caring for plants often primed gardeners to think more capaciously about their definitions of life.

The Garden Politic takes a historicist and materialist approach to account for the kinds of inspiration Americans found in studying and cultivating plants. This situated approach emphasizes the divergent ends to which cultivation was mobilized. As numerous scholars have already shown, for instance, the discourse of hybridity, subversive in some contexts, was also essential to the colonial project.[74] Hybridizing was a central strategy for adapting plants across climatic zones.[75] Likewise concepts like circulation, boundary crossing, sentience, and grafting had complex meanings. Gardens themselves could be spaces of freedom but were an equally effective technology of empire, naturalizing conquest under the guise of "improvement." And caring for plants did not necessarily translate into any specific political ethos but rather served as a site for understanding various political orientations. As anthropologists Aryn Martin, Natasha Myers, and Ana Viseu have pointed out, care is "a selective mode of attention: it circumscribes and cherishes some things, lives, or phenomena as objects. In the process it excludes others."[76]

Beyond anthropomorphism, one of the hard questions in plant studies is whether imagining plant life in more capacious terms can help change the fierce inequalities in our political systems, or whether it draws energy away from more urgent antiracist and decolonizing agendas. *The Garden Politic* shows how the nineteenth-century culture of plants played a fundamental role in the way that settlers imagined the nation. By connecting their observations about plants to their understanding of human politics, the cultural figures in this book remind us of the way in which capitalism and white supremacy have defined not simply humans, animals, and plants but the dynamic relationships between and among them.[77]

The Organization of the Book

The Garden Politic begins in the 1820s with the development of botanical education and the burgeoning of commercial nurseries and horticultural clubs. Phelps's *Familiar Lectures* and other similar botany books helped usher in an era of botanical study, and technologies like the steam engine and the Wardian case revolutionized conditions for transport. The following chapters trace the flourishing culture of plants across the middle decades of the century, a time when scientists and gardeners debated the capacities of plant life.

The book takes the East Coast, and particularly the Northeast, as its focus because horticulture and botany were institutionalized in New England earlier than anywhere else in the United States. The concentration of wealth and population and the history of transatlantic trade networks meant that, for at least the first half of the nineteenth century, horticultural developments coming from Europe tended to arrive on the East Coast and then spread west.[78] Yet as chapter 3 illustrates, to "see—New Englandly" is hardly a matter of regional or local view.[79] The circulation of plants into local gardens and greenhouses accustomed gardeners to think of connections across broader scales and vaster distances.

Each chapter in *The Garden Politic* explores a key theme in nineteenth-century U.S. horticulture. The first chapter shows how the idea of plant geography fostered a nationalist discourse and how authors adapted this discourse to accommodate plant circulation and transplantation. Figures like Thomas Jefferson and Hector St. John de Crevecoeur connected American identity to horticulture, and other writers working in the domestic genre popularized this connection, naturalizing the traffic of plants and white settlers. Lydia Sigourney promoted the idea that European settlement was organic through the logic of transplantation. Sentimental writer Lydia Maria Child, too, conceived of the American republic in biotic terms, and her shifting use of botanical language across her career is a particularly useful lens for tracking developments in American botanical culture as a settler-colonial project. In her early writings, *Hobomok* (1824) and *Evenings in New England* (1824), Child conceives of the body politic through an essentially imperial metaphor of grafting. By naturalizing the European presence in New England through the idea of transplantation, she constructs a botanical taxonomy in which citizenship is organized around an exclusionary horticultural sentiment. As her career goes on, however, Child revises her approach to suggest a more radical critique of U.S. racial politics. Departing from her early work in which a sentimental botanical nationalism upholds fairly rigid taxonomic coordinates, her last novel, *A Romance of the Republic* (1867), moves to deconstruct such taxonomies and adopts a more inclusive vision of a multiracial polity. At the same time, the novel remains committed to upholding the idea of the republic itself as an organic entity.

The next two chapters engage with two central concepts for understanding nineteenth-century horticulture: unpredictability and sentience. Chapter 2 demonstrates a middle-class cultural investment in the idea of cultivation as a master metaphor for control over society and attends to the subversive potential of plants that flouted human control over nature. The chapter explores the impact of these ideas by tracing them through the domestic fiction of the author and amateur gardener

Nathaniel Hawthorne, nephew of the esteemed horticulturalist Robert Manning. From the rosebush outside the prison door in *The Scarlet Letter* to Alice's posies in *The House of the Seven Gables* to lesser-known instances that pervade his writings, Hawthorne's work illustrates again and again the problem that plants pose to patriarchal authority. As horticulture popularized new material understandings of dispersal, dislocation, and circulation, Hawthorne's work tracks how the movement of plants could destabilize ever-expanding efforts to control natural environments. I attend closely to this mobility in following how Hawthorne's botanical images engage a European context of scientific knowledge, practices, and technologies. Hawthorne also drew on X. B. Saintine's *Picciola* to critique botanical infatuation as a means toward socialization. Though Hawthorne can be seen to fixate on and even support methods of social control, he embraces botanical mobility to undermine the idea that nature can be fully governed and that domestic order can be unproblematically based upon property ownership.

While botanical classification remained an important argument for the hierarchical order of the natural world, by the middle of the nineteenth century new experiments with plant sensitivity raised questions about plants' ability to feel and act deliberately. These experiments were widely publicized, inspiring public debate about the intelligence and agency of plants. Chapter 3 shows how emerging theories of plant agency unsettled popular thought about biological hierarchies. These ideas posed a linguistic as well as an ontological challenge to popular understandings of life. Alongside scientific and popular writing, poetry's formal possibilities provided a medium for exploring the limits of languages to characterize other forms of life. The chapter uses the plant practices and experimental verse of Emily Dickinson to reveal the radical potential of both new botanical theories of plant sentience and ordinary gardening practices. Like many of her contemporaries, Dickinson kept herbaria of pressed plants. The Dickinson herbaria contained not only samples of flowers in her garden but also specimens sent to her by friends and correspondents from Southern Europe, the Middle East, and Central Asia. Such correspondence became more common as missionaries from the United States traveled abroad, helping gardeners gain material comparisons between local flora and that of distant climes. Flower and garden imagery allowed Dickinson to link the scenery in Amherst to that of Santo Domingo and Circassia, showing the true geographic reach of the supposed "poet of Amherst" and signaling the similar range available to any gardener with a seed catalogue. Her poems draw on developing theories about plants' autonomous circulation and botanical sentience, and Dickinson demonstrates how plant vitality demands a recalibrated politics that takes into account the feeling nature of other forms of life. By exploring botanical feeling and mobility, Dickinson challenges hierarchical modes of sentience—from mineral to vegetable to animal to human—and asks us to consider a world in which other forms of life both exceed and redefine the parameters of the human.

The last two chapters consider popular horticulture in the context of plantation slavery. Chapter 4 examines abolitionist efforts to tie the horticultural fervor of the white middle class to an antislavery position, centering on one of the most popular literary abolitionists of the antebellum period: Harriet Beecher Stowe. Stowe was a zealous gardener whose environmental sensibility evolved as her career progressed, shaping her racial politics and abolitionist strategies. Stowe's passion for botany and plants led her to a sense of connection across the human and nonhuman world that collapsed hierarchy. "A garden seems to bring a man into confidential relations with all the forces of nature," she writes in 1855. As her career developed she drew on theories of plant vitality to refute the strict classification and cultivation practices associated with slavery, disrupting the logic used to segregate humans from each other and from the environment. Stowe celebrated botanical migration and a seamless transition between garden and wilderness as an alternative to the policed boundaries of plantation culture, and she ultimately endorsed biological diversity as essential to the viability of America's democratic project.

The last chapter continues to focus on the problem of plantation slavery and the afterlives of the plantation. Chapter 5 shows how Frederick Douglass emerged as the nation's most trenchant theorist of the garden politic in opposition to the slave-owning founding fathers who envisioned an agrarian republic. As an orchardist himself, Douglass saw clearly the contradictions between an emancipatory botanical rhetoric and the realities of a nation where cultivation was tied to slavery, especially after the failure of Reconstruction. Enslaved individuals who sowed and harvested cash crops were forced into intimate, monotonous relationships with plants like cotton, tobacco, sugar, indigo, and turpentine pines. From his home in Rochester, New York—a city with a high concentration of commercial nurseries and horticultural periodicals—Douglass's editorship of the *North Star* and *Frederick Douglass's Paper* brought him into constant contact with horticultural news. The chapter follows Douglass as he reveals how the burgeoning field of scientific agriculture and growing horticultural information networks ultimately served the plantation economy.

Finally, the conclusion considers renewed interest in plant intelligence and communication today as an expression of desire for political change. From podcasts to pop culture, from a Pulitzer-winning novel to Instagram influencers, the past ten years have seen a dramatic uptick in conversations about the sophistication of plant life. While many scientists remain wary about ascribing human qualities to plants, which they fear flattens the latter's complexity, many popular writers draw inspiration from recent scientific papers that characterize collective agency and communication in plants. In showing how forests, for instance, appear to function through altruism, selflessness, and multispecies cooperation, these authors hope plants might model a social safety net as a guiding political principle. Drawing on a number of recent literary works, including *Braiding Sweetgrass* by Potawatomi scientist and writer Robin Wall Kimmerer and *The Overstory* by Richard Powers, the conclusion reflects on what the contemporary plant turn might learn from nineteenth-century understandings of plant life.

The crises of the Anthropocene, including the COVID-19 pandemic, are also openings to rethink and redress the causes of environmental injustice. Solving such systemic issues is a matter of science and policy, but it is also a matter of imagination. As the writers here show, thinking of plants differently can attune people to different models of relation that are not built on the twinned exploitation of people and the more-than-human world. A return to nineteenth-century antecedents can remind us of the political possibilities of—and limits to—this long-standing form of environmental thought.

1

Botanical Nationalism

Sometime in the early nineteenth century a coconut was plucked from a tree in Point de Galle, Ceylon.[1] After consuming its meat and milk, a scrimshaw artist set to work carving the shell. The coconut, now sitting in the material culture collections at the Winterthur Museum, tells us all of this by way of a poem carved into its side:

> In Point DE Gall I Did Grow On A Tree
> So High A Black man Cut me Down
> A Sailor Did me Buy
> My Blood he Drank my flesh
> Did Eat my Raiment hove
> Away & here Remains
> My Ribbs & Trucks
> Until this Very
> DAY

Like an eighteenth-century "it" narrative, the coconut invites readers to imagine the circulation of an object.[2] What was its journey to "this Very DAY"? Who was the "Black man" who cut it down, and under what conditions? Who was the sailor of unnamed race who purchased the coconut and transformed it into an object for display and a medium for verse? And what stories could the coconut tell of the countless homes it has known in its lifetime, which spans across multiple continents and generations?

Unlike a typical "it-narrative," the coconut appears as a material testament of its own circulation, though there's no way of telling whether it is a reliable narrator. "Speaking" with words carved into its own flesh, the coconut sets loose a host of metonymic associations about the material conditions in which it was made. The language of flesh and blood underscores the body of the coconut, alluding to the vitality of the plant con-

Figure 1.1. Cup, unknown maker, England or United States, 1800–1875. Coconut shell, pewter and brass, 1965. 1836 A, B. Bequest of Henry Francis du Pont. Courtesy of Winterthur Museum.

sumed on the path to commodification and to the flesh-and-blood men whose labor transforms it, susceptible to similar kinds of extraction.

For the artist who made it, the coconut became a medium for connecting these questions to the imperialist forces that were colonizing the globe.[3] In addition to the poem, the coconut speaks in a set of symbolic carvings that represent some of the dynamics shaping its fate across several centuries. On one part of the coconut is carved an image of Britannia ensconced in a chariot and carrying the British flag. On another part is the massive eagle of the great seal of the United States grasping a banner with the words E Pluribus Unum. By the start of the nineteenth century Britain was heavily invested in extending its control over all of Ceylon, which had been partially subject to Portuguese and then Dutch rule in the centuries prior to Britain's imperial conquest. Britain had claimed parts of the island from the Dutch in 1796 and took full control of the island by 1818. By contrast, U.S. involvement in Ceylon in the early nineteenth century was limited to commercial interests—including whaling vessels and spice traders—and a growing number of missionaries.

The coconut began its decorated life in an era when both Britain and America devoted considerable energy to economic botany, or the study of plants deemed important to human use. This study was tied to imperial efforts to search for and lay claim to new and valuable plants growing around the globe. From the eighteenth century onward, bioprospecting was, in the words of Londa Schiebinger, a "big science and big business" that radically transformed the global landscape and made vegetation into a key resource for facilitating economic and imperial expansion.[4] As Schiebinger and others have shown, political economists across Europe advocated natural knowledge as a means to national wealth and power.[5] Economic botany is usually understood as the province of politicians and scientists on expeditions, but sailors, scrimshaw artists, literary authors, and educators likewise stressed the national significance of plants in both economic and symbolic terms.

In New England, economic botany also became a way that domestic gardeners and children in New England were encouraged to imagine their own homes. In one 1824 book aimed at young members of the republic, in fact, the coconut appears as an example of a plant from the tropics with many domestic uses, its utility shoring up an argument about the value of a botanical education. "Do you believe that there is a tree in Mexico, which yields water, wine, vinegar, oil, milk, honey, wax, thread, and needles?" an aunt asks her niece in Lydia Maria Child's *Evenings in New England*. Not only do the various parts of the coconut tree supply all these commodities, the aunt explains, but bowls, baskets, brooms, nets, mats, sacks, and other utensils besides, as well as an antidote to certain poisons.[6] In the same chapter the aunt ties a number of other common household objects to plants with origins in other parts of the globe. Approaching trees from the perspective of domestic economy—that is, with an interest in their domestic usefulness and the ubiquity of household objects produced out of their raw materials—the aunt emphasizes that perhaps the most awesome power of trees is their utility.

As the collection as a whole makes clear, and as we might expect from the author of *The American Frugal Housewife* (1829), botanical tropes were popular for allegorical purposes, but they also mattered from the perspective of domestic management. Most critical studies of domestic gardening activities in America have attributed the reasons for its growing popularity in the early decades of the nineteenth century to the characterological benefits purported to accompany botanical study.

And indeed, gardening and botany were popularly associated with the development of the morality, health, and organization of the individual and, by extension, the nation. From this perspective, botany served women well because it allowed them to derive lessons from the natural world while remaining within the parameters of the domestic sphere.[7] But *Evenings in New England* links the home economy project and the educational value of botanical instruction to the global circulation of plant goods.

Child was an antislavery activist, early feminist, accomplished author, and domestic educator, but she was also a keen gardener savvy about the role of horticulture in developing a national economy.[8] Her first children's book reveals how extensively domestic home gardening engaged with the geopolitical circulation of plants that, for instance, turns "sap, which flows from several trees in the East Indies and South America" into the piece of India rubber the aunt holds in her hand (19). Child's first children's book turns what scholars call "bioprospecting" into a bed-time tale.

This chapter considers how botanical nationalism took shape in pedagogical and domestic literary works in early nineteenth-century America. The ideas of "native plants" and "American soil" served as the basis for a pervasive botanical nationalism, a conception of the republic in distinctively biotic terms. The power of this formulation lay in its flexibility; rooted into the ground, plants provided a material basis for claims to native distinctiveness. At the same time, the circulation of plants through commercial and personal channels led to the influx of new species, whose successful transplantation could stand as a harbinger of successful human adaptation. As Europeans continued to immigrate to America in the nineteenth century, they brought plants with them, and the adaptation of one could serve as a potent metaphor for the Americanization of the other. The circulation and adaptation of plants, in other words, was another potent dimension of botanical nationalism.

Child is an instructive figure in this sense, as someone who self-consciously crafted herself as an American author and who drew heavily on her knowledge of plant culture to depict domestic life and domestic plots.[9] This chapter draws on a range of early nineteenth-century writing, using Child's work in particular to illustrate the number of forms this botanical national identity could take. In *Evenings in New England*,

Child shows how economic botany and domestic economy overlapped in material ways, spurring biotic understandings of the nation that accommodated the transnational web of scientific, economic, and political plant circulation. Child's first novel, *Hobomok*, appeared the same year she published *Evenings*, but it emphasizes the idea of transplantation to a settler imaginary of "America." Thus she engages the idea of the botanical nation in different ways—the one to demonstrate the economic value of plants to domestic visions of America, and the other to naturalize settler colonialism in symbolic terms. Later in her career Child turned once again to botanical language, but to critique rather than shore up nationalist rhetoric. By the time Child wrote *A Romance of the Republic* (1867) after a long career as an activist, she perceived neat botanical ordering as a tool for patriarchal and imperial power, and thus employed flowers not as a way of naturalizing national identity in a celebratory mode but as a way of diagnosing the taxonomic disorder produced by slavery. The differences between these three books reveal how botanical literary reference was not a monolithic discourse but rather part of a capacious and unstable category of symbols that underpinned diverse constructions of the nineteenth-century body politic.

Economic Botany Lessons in the Early Republic

When botany started to burgeon as a discipline in America during the 1820s, it was often promoted on moralistic grounds, especially in terms of its appropriateness for women. In *Familiar Lectures on Botany* Almira Lincoln Phelps elaborates on how the science helped inculcate a number of qualities valued in the young republic. It might "illustrate the most logical divisions of Science, the deepest principles of Physiology, and the benevolence of God" and promote "heath and cheerfulness"; it might also, panacea-like, "serve to interest and quicken the dull intellects of some pupils, to arrest the fugitive attention of others, and relax the minds of the over-studious."[10] Catharine Beecher would later echo similar sentiments in praising botany for the "formation of habits of investigation, of correct reasoning, of regular system, of accurate analysis, and of vigorous mental action."[11] Botany was an important world-ordering system with aesthetic appeal, and as Phelps writes in the first edition of her popular text, "The analysis of even a few flowers cannot fail of suggesting thoughts of the beauty of a system which so curiously identifies the different plants described by botanists, and points to each individual of the vegetable family the place it must occupy."[12] Phelps believed that an ordered understanding of plants gained through botanical science could service national interests. The imperative reflected in Phelps's logic will be revisited in subsequent chapters, for botany's love of system generally aligned with a hierarchical sense of society.

Child's *Evenings in New England* likewise strives to educate a young audience on the benefits of botanical education, though her approach ranges from the scientific to the allegorical. A section titled "Riddling Forest" is a list of botanical wordplays. The enigmatic prompts range from "What tree moves backward?" (crab) to "What tree has for ages withstood the fury of the oceans?" (beech).[13] In a more heuristic section, "The Uneasy Oak" relates the cautionary tale of one disgruntled tree who begs to be transplanted to a city and dies shortly thereafter: a parable about being content with one's lot in life. And "Adventures of a Dandelion" relates an allegory about class consciousness and economy told from the perspective of a dandelion and transcribed by a snail amanuensis after the dandelion is pulled "limb from limb" by a prospecting naturalist (38).

Child's interest in plants extended beyond moral education, however, and the chapters of *Evenings in New England* that emphasize the importance of botanical instruction do so to emphasize that such knowledge carries tangible domestic benefits. By underscoring the utility of trees in this context, Child was participating in an ongoing conversation about the importance of biotic matter to the country's economy. Naturalists in America understood their work as matter of national significance. Catalogs chronicling the botanical species of North America proliferated in the first several decades of the nineteenth century and attempted, like Jefferson's earlier *Notes on the State of Virginia* (1785), to quantify and qualitatively describe the continent's natural resources. To facilitate this task, American naturalists appealed to citizens to send them interesting specimens for evaluation.[14] In return, they promised to provide these correspondents with knowledge about their submission that might, as Andrew J. Lewis has argued, help them turn a profit by capitalizing upon some as-of-yet "unidentified" but valuable plant, mineral, or animal.[15] These naturalists emphasized that the value of botanical practice was collective as well as individual—the country stood to benefit from the promise of economic independence that came with commodifying nature.[16]

But the alignment of nature and nation went deeper than a simple economic nationalism built around treating the landscape as natural resource. Writer Hector St. John de Crevecoeur contrasts the "espaliers, plashed hedges and trees dwarfed into pigmies" to be found in Europe with America's wild cherries, which are to be celebrated "as nature forms them here, in all her unconfined vigour, in all the amplitude of their extending limbs and spread ramifications." Visitors, he notes, will no doubt observe that "we are possessed with strong vegetative embryos."[17] Likewise, Jefferson's *Notes on the State of Virginia* adumbrates the variety and vigor of the country's natural resources to legitimate the country's democratic ideals. Both men relied upon a trope that would become widely prevalent with the rise of self-consciously national literature: the uniqueness of the American soil and the plants that grew there. In 1821, the *North American Review* proclaimed the need for a more extensive national botanical program, stressing the importance of the "difference and affinity" of plants found in North America and Europe. This knowledge would help facilitate trade with the continent, but also, the author implies, provide a botanical basis for establishing national distinctiveness.[18]

As much as American botanists sought to establish a relationship between natural and national distinctiveness, they also facilitated an extensive cosmopolitan network of natural history exchange. Benjamin Smith Barton, for instance, a renowned Philadelphia botanist and professor at the University of Pennsylvania, spent time in both England and France and corresponded with Russian naturalists to the extent that he was inducted into the Imperial Society of Naturalists at Moscow. Thomas Nuttall, a prominent naturalist and émigré from England, likewise cited a large cast of influential botanists in the preface to his 1818 *Genera of North American Plants*, including a Frenchman (de Jussieu) and two Germans (Christian Sprengel, Frederick Pursh). Determining that "in compliance with the public to whom it is addressed, an uniform language appeared necessary," Nuttall apologizes in his preface for writing merely in English and notes that "the great plan of natural affinities, sublime, and extensive, eludes the arrogance of solitary individuals, and requires the concert of every Botanist and the exploration of every country towards its completion."[19] His introductory remarks suggest the extent to which botanists participated in a collaborative transnational dialogue even as they sought to shore up the interests of their own governments.

This fact is evident in Almira Phelps's career as well. After first publishing *Familiar Lectures on Botany* in 1829, Phelps continued to revise it over the subsequent decades, and it was reprinted over twenty-eight times.[20] Her preface to the 1854 edition both alludes to the territorial conquest that enables her to add the "new genera and species of *Southern* and *Western* plants, as well as those of more *Northern latitudes*" and expresses her debt to European publications: "The Author of this work, in its preparation more than twenty years since, availed herself of the most valuable foreign works, consulting English books less than those of the French and German school of Botany, so that in reality much that [English naturalist John] Lindley brings forward as of 'foreign origin,' had previous found a place in this work."[21] In establishing a continental genealogy for her original text, Phelps echoes Nuttall's conception of botany as a collaborative endeavor. And by dismissing Lindley's claim to originality, Phelps asserts both her intellectual priority in this regard and her ongoing dialogue with European botanists. Phelps's comments demonstrate how botany, like many sciences in the Romantic period, involved both transnational collaboration and national rivalries.[22]

Literary authors and educators likewise stressed the national significance of plants in both economic and symbolic terms and recognized national botanical interests as reaching beyond the borders of the nation. In depicting the world's trees and plants as potential resources for the young nation, *Evenings in New England* acknowledges botany as a global science link to national prosperity. In "TREES," the aunt's musings cover a wide terrestrial range, pairing place with product. In the South Sea Islands, she notes, bark is used to make clothing; in the Ladrone Islands a certain tree fruit is used to make bread; in Japan there is a tree that makes fine paper; and in Jamaica there is a "lace tree" that, while producing none of the fine cloth, bears its resemblance. The aunt's list of use value goes on, linking coffee cultivation to "Arabia, Persia, the East Indies, the Isle of Bourbon, and some parts of America" and camphor to a species of laurel from China. She places the growing range for Gamboge (a "useful medicine and paint") likewise within specific geographical regions: "Ceylon, Siam, and Cochin China" (22). As the encyclopedic style of the vignette attests, the study of botany produces a global map of trees that can be turned into commodities.

The political implications of commodity cultivation at home were likewise not lost on Child. In 1839 she moved with her husband David Lee Child to Northampton, Massachusetts, so that he could pursue a business in beet sugar. Beets, which could be grown in colder latitudes than cane, held out the promise of increased sugar production in the United States without high import costs or slave labor. Although her husband's venture famously failed, the attempt revealed a working commitment to changes in the U.S. agricultural system, with an attendant belief in the political power of such a shift.

Child was not alone in explicitly appraising the material value of local biota. Other popular writers of the period explicitly considered the significance of local medicinal knowledge about plants to the health of the republic. The widely read poet Lydia Sigourney, for instance, likewise understood the material value of flowers in addition to their sentimental significance. Sigourney's father was a gardener for a wealthy family in Norwich, Connecticut, and Sigourney encouraged her readers in the scientific study of plants. In her popular conduct book *Letters to Young Ladies* (1833), Sigourney praises first the moral benefits of tending flowers and then the "practical import" of studying them, hailing their "coloring matter" and "healthful influences" as principal benefits of botany.[23] An advocate for Native American rights even as she adopted an assimilationist stance, Sigourney praised the botanical expertise of Indigenous women. Suggesting that readers might look for inspiration to "female aborigines of our country" who "were distinguished by an extensive acquaintance with the medicinal properties of plants and roots, which enabled them, both in peace and war, to be the healers of their tribe," Sigourney promoted the idea that medicinal knowledge of plants was a worthwhile enterprise. While deferring to the expertise of male physicians, Sigourney nevertheless encouraged her readers that "sometimes" one's own knowledge of herbal infusions is warranted, falling as it does within "a legitimate branch of that nursing-kindness, which seems interwoven with woman's nature."[24] If plants were to be admired for their beauty, Sigourney suggests, they were also to be appreciated for their ability to heal.

Nor was Child alone in promoting a juvenile literacy in economic botany. Francis Lister Hawks, southern Episcopal clergyman and North Carolina politician, wrote a number of instructional books for children under the pseudonym Lilly Lambert. His 1834 *The American Forest: or, Uncle Philip's Conversations with the Children about the Trees of America* at times bears striking resemblance in tone and objective to Child's *Evenings*. Each chapter, structured around a different "conversation," stresses the use value of trees found in different parts of the country. But the first conversation starts abroad, assessing the utility of the coconut to Pacific Islanders before turning back to uses of American pines. As the children ask questions, Uncle Philip's answers steer them to think in terms of a global market, even as he insists that "American children ought to know something of American trees" so as to prevent deforestation.[25] Amid discussions of how to make canoes out of birchbark and sugar out of maples, one conversation contains an explicitly identified "Short Lesson in Political Economy" in which Uncle Philip teaches the children gathered round to consider labor costs as well as climate to evaluate what crops might grow.[26] During the latter part of one conversation, one of the children proposes to Uncle Philip that the United States might be able to grow everything within the various climates of the country, musing that commodities like sugar, olives, and tea might be grown in the South. Uncle Philip responds with an economic assessment, noting that it might be physically possible while also unprofitable. Saying nothing about slavery, the uncle mentions that wage labor in other "thickly inhabited" countries costs less.[27] And the whole of the book sees Uncle Philip fielding questions about whether it would be more economical to grow different crops, like cork or coffee, in the South, or to import them from elsewhere.

In Child's *Evenings*, the aunt's overarching argument for the study of plants includes a similar emphasis on utility. This aspect of botanical study was frequently discussed in the popular press, and many articles focusing on materials sourced from trees abroad described the process of harvest and shipment with great interest. Rubber was one such commodity that elicited curiosity as to its means of production and circulation, as the aunt's mention in *Evenings* attests.[28] An 1849 series on "Wonderful Trees" in *Robert Merry's Museum* declares that "the India rubber tree (*ficus elastic*) affords a product of such various and still multiplying use, that to be cut off from this article of commerce would be a serious loss to the accommodations of civilized life." After detailing the countries where rubber trees grow and the date that their use value was discerned in Europe, the article describes how India rubber is made by extracting the sap, and how it is stored for shipment. Once the commodity is identified, the tree is left behind and the focus becomes the plethora of uses for the product. Reading almost like an advertisement, the article notes that "India rubber is the companion, in one form or another, of all ages, from infancy to manhood. Babies suck and bite India rubber rings. . . . When those babies have become boys, they play with India rubber balls, or make miniature ferry boats. . . . When still older, they supply themselves with a water-proof suit, dress themselves from head to foot in caoutchoc [*sic*], and with a tent and boat of the same article, are off for the golden shores of California."[29] The passage aligns human growth not with the growth of the rubber tree but with a growing use of its main product, put to diverse uses. In other words, it narrates the tree's identification as a raw material for commodity production and not as an organic metaphor. Articles like this are abundant in the antebellum period, and they normalize extraction or removal by linking it to the usefulness of the end product, even as they marvel at the seemingly endless possibilities of commodification.

Just as the commodities that these trees produce circulate around the globe, the discussion between niece and aunt in *Evenings* makes clear that the trees themselves are circulating commodities that support national pride. Horticultural societies frequently conceived of their efforts in nationalistic terms, seeking to demonstrate that their tree collections were on par with European institutions, and they often conceived of America's horticultural project as part of an ancient horticultural lineage that reached back to famous empires. In marking the one-year anniversary of the Massachusetts Horticultural Society in 1829, its president proclaimed, "Most of our common fruits, flowers, and oleraceous vegetables were collected by the Greeks and Romans from Egypt, Asia, and other distant climes, and successively extending over Western Europe, finally reached this country."[30] Such narratives situate American horticultural projects within a narrative of westward global development, a lineage that Child similarly rehearses in *Evenings*. After the aunt claims the African and the Middle Eastern provenance of the acacia tree, her niece notes that such trees exist in American gardens. Assenting, her aunt moves on to discuss foreign plants in America. "You will often see those standing side by side," she observes, "which originated in the re-motest corners of the earth." A genealogy of foreign plants in New England follows this observation and is the means by which the aunt situates the United States within a long historical narrative. The ubiquitous horse chestnut, for instance, "which shows its proud form and beautiful foli-age amid the bustle of Boston, New York, and other southern cities,—or blooms fresh and green in some elegant garden in their vicinity, came from the famous Mount Pindus in Arcadia—that country which poets have always represented as the most beautiful in the world."[31] This heri-tage, the aunt suggests, links bustling American cities to a "proud" clas-sical tradition represented not by the erection of triumphal arches or ionic columns but through plant networks. In this way Child grafts her own historical associations onto the American landscape through bo-tanical transplants.

Evenings in New England reveals Child's sensitivity to applied botany—the consideration of a plant's usefulness as a resource for human enterprises—as well as her awareness of the bioprospecting and collection activity responsible for some of America's most prevalent plant species in the early nineteenth century. Child presents this activity in a wholly positive light throughout the collection of vignettes; sourc-ing plants is a matter of national pride, and the tone of the exchange is one of patriotism, wonder, and education. The idea of transplanting tra-ditions, especially from European and classical sources, can be seen in the way that American authors often paid homage to contemporaneous literary productions in Europe, and the extent to which the American literary marketplace was early on saturated with texts from abroad.[32] The positive references to botanical relocation, however, seek an authen-ticity derived from the very material transpositions that could literally continue to grow in the New World. Transplantation carried with it the promise of an organic connection to a classical past rooted in American soil. In this sense, it served conflicting impulses toward establishing a strong European heritage and demonstrating national distinctiveness. Child sought to reconcile these conflicting impulses in her first novel, *Hobomok*, published the same year as *Evenings in New England* but with, as we will see, a very different focus on horticultural sentiment.

National Botanical Sentiments

Child began her career at a moment when the nascent American literary establishment was clamoring for more explicitly American works. Writers who responded to this call, such as her contemporary James Fenimore Cooper, frequently turned to the wilderness as a potent conceit for something essential and naturally "American." *The Last of the Mohicans*, for instance, published in 1826, opens with this line: "It was a feature peculiar to the colonial wars of North America, that the toils and dangers of the wilderness were to be encountered before the adverse hosts could meet." After foregrounding the wilderness from the start, the rest of the first paragraph lays out the challenges facing fighters in the Seven Years' War—stream rapids, difficult mountain passes—before they even meet their adversaries. And nineteenth-century critics calling for a distinctively American voice, as Karen Kilcup has shown, "concluded that American poetry was most original and most genuine when it dealt with the untamed landscape or its putatively wild inhabitants, indigenous peoples—safely ensconced in well-ordered verses."[33] In short, consensus largely held that the uncultivated—in nature or person—represented the genuinely American. But if poets sought to harness wild nature in regular verse, they also sought to Americanize a nation of immigrants through metaphors of cultivation and transplantation. This was especially apparent in the writings of several popular sentimental writers.

Lydia Sigourney, the first American to make a viable career out of her poetry, understood the potency of other kinds of organic metaphors in this regard. Her sentimental depictions of floral culture are well known. Less acknowledged are the sophisticated ways that she turns to floral transplantation to make white European immigration into an organic, natural phenomenon, one in keeping with the growing popularity of nursery culture and celebration of the ability to grow exotics from abroad at home. The ability to take root and to thrive in a new climate was a natural and racial argument for fit.

A number of Sigourney's poems unpack the significance of botanical language to establishing an immigrant's rightful place on new soil. In her popular collection *The Voice of Flowers* (1846), Sigourney included poems that describe the transport of plants from Europe to America.

Both "The Rose-Geranium, Companion of a Voyage" and "The Emigrant Daisy" describe the significance of transplanted flowers to the human speaker's adjustment to a new continent. In "The Rose-Geranium," the speaker narrates the stresses put upon a flower while shipped across the ocean: "It grieves me sore to see thy leaflets fade, / Wearing the plague-spot of ocean spray, / And know what trouble I for thee have made, / Who bore thee from thy native haunt away." Yet the speaker's emotional need for the flower justifies the transplant, and is readily apparent in the closing lines of the poem; the companionate geranium provides a tangible link to the life left behind: "Though, in thy life, I seem to hold the chain / Of home and its delights, here on the pathless / main."[34] The flower becomes a metonym for home on a distant shore, one that can be rooted in new soil.

"The Emigrant Daisy" likewise addresses the theme of transplan-tation, but it carries a more explicitly political message. The poem's speaker digs up a daisy from England to carry to America, whispering "in its infant ear / That it should cross the sea, / A cherished emigrant, and share / A western home with me."[35] As in the prior poem, the plant has a domestic role to comfort the speaker in a new place. But a foot-note clarifies that the daisy was taken from the exact location where the Magna Carta was signed in 1215, and the next stanza aligns the daisy with the charter, noting that it survived the ocean journey "As if old Magna Charta's [sic] soul / Inspired its fragile form." Upon arriving in America the speaker sows "the choicest seed" "within my garden plat" and waits to see if this garden experiment will succeed. The political cli-mate should assure success, the speaker reasons, given that the "Magna Charta's [sic] spirit lives / In even the lowliest cell" in America, allowing that the "simplest daisy may unfold / From scorn and danger freed."[36] The poem's movement between the domestic and the political makes the home garden a space where biotic experiments are freighted with political weight.

Sarah Josepha Hale, writer and editor of *Ladies Magazine* and the even more influential *Godey's Lady's Book*, similarly understood flowers as situated within a transnational dynamic even as she used them to ad-vocate for the existence of distinctively American sentiments. Her popu-lar engraved *Flora's Interpreter; or, The American Book of Flowers and Sentiments* (1832) links the project of learning botanical names with the project of learning "American" sentiments. Filled with an alphabetical list of popular flowers, the book provides Linnaean classification for each flower, a description of its native range, a descriptor of the emotion with which the flower is identified, and a pair of poetic excerpts that evoke the flower and the correlated sentiment. Putting such a book together, Hale acknowledges, is not particularly distinctive. An increasing number of "language of flowers" books in England and America meant that the combination of a sentimental and scientific presentation would have already likely been familiar to many of her readers. Yet Hale stresses that what is new about her material is "the introduction of American sentiments."[37] While she often turns to European writers to serve as "authorities for the signification affixed to each flower," Hale takes pains to stress that "in the *sentiment* which the flower when presented is intended to convey, I have preferred, exclusively, extracts from American poets." The same was not necessarily true for the flowers themselves. Writing at a moment when print culture was, as Trish Loughran puts it, "without rival, American nationalism's preferred techno-mythology," Hale vaunts American poetry by associating its carefully selected sentiments with plants from around the globe that might be cultivated at home.[38]

Hale emphasizes that the flowers she chose were accessible to American gardeners even if not originally native to North America. Listing the continental range of each plant, Hale rationalizes that "a knowledge of the locality of the plant would . . . assist us to judge somewhat of its character and adaptation to our gardens or green houses."[39] The dandelion, for instance, is "indigenous to Europe, but naturalized in America."[40] The geranium's range is extensive, found in Europe, America, and Africa, though the African species is "much the most beautiful and most cultivated."[41] Attaching American sentiment to plant species from different continents, Hale points to no boundaries in terms of what a gardener, equipped with financial and horticultural resources, might acquire.

If Hale and others encouraged the attachment of American sentiments to global flora, many nonetheless understood native plants in material terms as vital to health. Medicinal applications represented one of the most important use values of plants in the nineteenth century, and local herbal remedies were prized for their ability to heal ailments particular to the climate. While physicians often dismissed alternative forms of medicine, such as the botanically based Thomsonianism, as quackery, in reality herbal remedies played a significant part in an array of healing arts.[42] Enslaved Africans brought botanical knowledge with them across the Middle Passage, and Indigenous Americans likewise had a long history of drawing upon a robust herbal pharmacopeia. As Sharla Fett, Susan Scott Parrish, Londa Schiebinger, and others have shown, colonists in the Americas relied heavily upon the botanical knowledge of Native Americans and enslaved Africans. Early English colonists feared what the hot climates of the West Indies and southern United States would do to their temperate physiologies.[43] Believing in a correlation between local environment, disease, and health, they asked physicians to use endemic plants as curatives and sought knowledge from the very people they subjugated.[44] Moreover, westward expansionist efforts relied upon knowledge of flora and fauna, what Conevery Bolton Valencius has called "a formal tool" of American expansion.[45]

Valencius has described how "environment-specific illness and environment-specific treatment required practitioners to be *of* their regions in powerful and lasting ways."[46] White herbalists frequently sought to advertise their own local botanical expertise by association with Indigenous persons. During the same period that the 1830 Indian Removal Act went into effect, stripping the majority of Native Americans in the Southeast of their land, a small number of alternative medical practitioners used the epithet "Indian Physician" to distinguish their botanical approach to healing from other methods such as bloodletting. J. W. Cooper, who practiced near Lancaster, Pennsylvania, called his 1833 treatise on the topic *The Experienced Botanist or Indian Physician, Being a New System of Practice Founded on Botany*. Significantly, he makes no specific mention of and grants no credit to any Indigenous people for any of the information within the book.

Advertised "for the use of families and practitioners," Cooper's book rationalizes that curatives are to be found in local environs. He writes in the preface of "the means of health scattered so profusely around us, in almost every field and forest, placed, as it were, within the reach of every hand that will deign to accept it" and subsequently asks, "Is it credible that diseases, peculiar to our climate and country, can find no remedies nearer than foreign countries? Is it credible that the thousands of vegetables that beautify and perfume our fields and groves, have no valuable use in relation to the health and comfort of man?"[47] Cooper adopts a nativist philosophy when it comes to curatives; the remedies that Americans need are most likely already in their backyards, or at least can be found growing somewhere within the country. Continuing, he reasons that "the more the medical properties of vegetables have been explored, the more they have been found to furnish supplies for the nourishment and health of the animated part of Creation. Exotic vegetables may perhaps prove useful, especially if naturalized by cultivation in our own soil, in our own climate; but it is at least probable, if not certain, that our own native plants are abundantly sufficient to answer all the medical demands of our country."[48] The answer to what ails, Cooper suggests, can be found growing nearby. Beyond invoking Native American expertise in his title to lend credibility to his claims, Cooper includes numerous "certificates" after the preface, firsthand accounts that praise the efficacy of his treatment method. According to his boosters, Cooper's plants work nothing short of miracles.

Just as local plants were believed essential to a healthy polity, horticulturalists also stressed the role of cultivation in the process of Americanization. The polemical English writer William Cobbett penned one of the earliest treatises about gardening in the young republic. In *The American Gardener* (1819), Cobbett tailored his writing explicitly for a national audience, extolling the many virtues of gardening in this political and geographic climate. Aiming to increase the practice of gardening within the republic, Cobbett emphasized its potential for profit and conduciveness to health. But in closing his preface, Cobbett also praises the endeavor because it "is indulged *at home*. It tends to make home pleasant; and to endear to us the spot on which it is our lot to live."[49] Gardening, in short, breeds attachment and endearment to place, turning a "spot on which it is our lot to live" into a home, and a pleasant home at that.[50]

Andrew Jackson Downing, one of the most famous landscape gardeners in the nineteenth-century United States, likewise stressed the distinctive role of horticulture in fostering national identity. Downing's 1841 *Treatise* articulates the need for an American treatment of landscape gardening, noting that "even those who are familiar with foreign works on the subject in question, labour under many obstacles in practice, which grow out of the difference in our soil and climate, or our social and political position." Identifying Americans as "a people descended from the English stock," Downing nonetheless describes "our peculiar position, in a new world that required a population full of enterprise and energy to subdue and improve its vast territory."[51] For Downing, national character is the driving force in horticultural transformation.

Key to Downing's vision is that this character takes shape at home, in sentimental attachment to a particular place. Domestic feeling is a key aspect of Downing's vision of a horticultural America. "The love of country," Downing writes, "is inseparably connected with the *love of home.*" It begins with "innate feeling, out of which grows a strong attachment to natal soil."[52] By improving the home through "sylvan and floral collections," the "country gentleman of leisure," Downing's main audience, "improves the taste, and adds loveliness to the country at large."[53] Metonymically conflating home with nation in a familiar manner, Downing unfurls a vision of horticultural colonization: "The taste and the treasures, gradually, but certainly, creep beyond the nominal boundaries of the estate, and reappear in the pot of flowers in the window, or the luxuriant, blossoming vines which clamber over the porch of the humblest cottage by the way side."[54] Downing's vision of "creep" here suggests how a gardener's good taste might naturally spread across the nation—a settler fantasy of conquest unfolding organically through steady horticultural expansion.

Taken together, what these disparate visions share is a strong sense of the relationship between home and country as navigated through sentimental cultivation. Home is rendered as a material and sentimental relationship with the surrounding flora and fauna. Moreover, this flora does not necessarily have to be local to be associated with home and country. Both local and transplanted biotic material could serve as the basis for national sentiment. Lydia Maria Child's *Hobomok* is a particularly apt example of this kind of vision, using botanical imagery to argue for the naturalness of a pluralistic settler-colonial society.

Hobomok, set in the early colonial period, revolves around a Puritan girl named Mary Conant who has come from England with her parents to settle in Massachusetts. Her father disapproves of her love interest, a man named Charles Brown, because of his Episcopalian faith, and Charles leaves the colony for England. His ship crashes on the way there, and in grief and rebellion Mary flees her father's house and marries a Pequot man named Hobomok. They have a son together and share a peaceful domestic existence until Charles returns, having in fact survived the shipwreck. Hobomok relinquishes custody of his son and departs Massachusetts, leaving Mary to reunite with Brown and their son to assimilate into whiteness.

In keeping with the novel's attempt to articulate a distinctively American identity, *Hobomok* leans heavily on the trope of botanical hybridity, depicting the American republic as a hybrid between Native American and English influences. In this respect Child's first novel is exemplary of what becomes a pervasive antebellum cultural formation: naturalizing national identity through a bloom narrative that links human sexuality to plant culture.[55] Floral tropes map onto the plot in a fairly stable way, triangulating espoused values, thriving plants, and national identity. *Hobomok* has long been regarded as a novel about the role of women in challenging patriarchal authority and mitigating the tensions between conflicting systems of belief. Yet the sentimental botanical nationalism that Child constructs had enormous cultural power that adumbrates and extends beyond gender politics. Considered from a racial and national vantage point, the idea of the graft, in particular, served the ideology of what proved a limited kind of pluralism and provided a model for paying homage to English culture while simultaneously claiming national distinction. More significantly, it naturalized European claims to the landscape, displacing Native Americans.

Child, of course, was hardly original in using botanical science to represent sexuality and courtship.[56] Linnaean taxonomy relies upon the number of reproductive organs for classification, and the organization of plant classes drew easy analogies to human anatomy in both scientific and literary circles. What makes *Hobomok* notable, however, is how Child explicitly uses the literary conventions that associated female sexuality with floral blooming—what Amy King calls a "botanical vernacular"—to correlate the success of the American political project with a feminine botanical metaphor.[57] Child set out in *Hobomok* to represent the fledgling nation's exceptional identity through botanical language and logic. The plot conceptualizes how the young republic might reconcile different religious and cultural practices—Puritan and Episcopalian, Native American and English—and how these same practices might constitute American identity.[58] Mary Conant, the protagonist whose health, happiness, and vitality form a proxy for the republic as a whole, is described by her parents as their "youngest little blooming fairy."[59] Charles Brown, Mary's first love interest, is deemed by her father to be one of the "strange slips which are set upon our pleasant plants" (10) because of his Episcopal faith. For all Mr. Conant's scorn, Child presents the act of this particular graft as an ideal, for by the end of the story Brown's Episcopalianism is successfully integrated within the Puritan community.

In the antebellum period, grafting was popular as a horticultural technique for several reasons: it offered a highly successful means of propagating new varieties and stimulated the plant's growth. Graft and stock share sap but retain their distinct characters. Or as *The New England Farmer, and Horticultural Register* put it, "The scion, bud, or inarched shoot, is endowed with the power of drawing or forming from the stock that peculiar kind of nourishment which is adapted to its nature, and . . . the specific characters of the engrafted plant remain unchanged, although its qualities may be partially affected."[60] The graft was appealing not only as a horticultural technique but as a trope for identity, as when Hector St. John de Crevecoeur, in debating his American and English affiliations in *Letters from an American Farmer* (1782), notes that English national sentiment was "grafted upon the first rudiments of my education."[61] As a metaphor, the grafted slip embodies unity without destroying difference, for the stock and slip grow together while yet retaining distinct characteristics.

Charles's eventual naturalization into the Puritan proto-national fold is largely accomplished by the way the floral imagery bridges the romance plot to the national narrative. Over the course of the novel, Mary's mother, Mrs. Conant, who in emigrating from England with her husband had "left a path all blooming with roses and verdure, and cheerfully followed [her husband's] solitary track" (16), begins "drooping" (108).[62] Mary, who has likewise been transplanted from the "blooming gardens of good old England" (48) where her grandfather had raised her "with more than tenderness, like some fair and slender blossom in his gardens" (78), must take to the new environment. Her initial struggle to do so, a struggle that underpins the novel, is conceived through similar botanical language: "What was she now? A lily weighed down by the pitiless pelting of the storm, a violet shedding its soft, rich perfume on bleakness and desolation; a plant which had been fostered and cherished . . . removed at once from the hot-house to the desert" (79). Mary's floral associations do more than identify her as a young woman of marriageable age; they characterize her displacement from England and subsequent acculturation to her new environment.

The eventual success of Mary's transplantation, which eventually accommodates Brown's incorporation as well, is facilitated by her marriage to Hobomok. As Priscilla Wald notes, Mary becomes Americanized through this "ill-advised" marriage to the chief of a neighboring tribe.[63] More specifically, Hobomok's strong association with American nature helps Mary put down permanent roots. Once they have a son, an occurrence that Child treats with equanimity, Hobomok's presence is no longer essential to her process of naturalization. With speed, then, Brown—presumed drowned but not dead after all!—returns to the settlement and Hobomok leaves his wife and son so that Mary can be with her white lover. By the novel's close, Mary and Hobomok's son goes to finish his studies in England, where "his father was seldom spoken of, and by degrees his Indian appellation was silently omitted" (150). This ending has rightly drawn intense critique for uncritically excluding Hobomok from his paternal rights and, by extension of the sentimental logic of the text, from his place in the republic.[64] But Child's horticultural imagery, as we shall see, displays more ambivalence about his departure.

One of *Hobomok*'s central conceits is that love is a governing force of social relations in the new republic, rendering sentiment along with sexuality a matter of plant blooming. Sentiment, as Laura Mielke, Kyla Schuller, and Ezra Tawil point out, was essential to the construction of raced and gendered national subjectivity in the antebellum period.[65] Mary's grief over Charles's drowning precipitates her rebellion against her father and marriage with Hobomok. And Hobomok's love for Mary likewise governs the novel's other major plot turn: Hobomok's decision to leave Mary and their son when he discovers that Charles not only is alive but has returned to the colony. Attached to love's narrative force is Child's characterization of love as the "humbler blossoms of the heart" (47). Hobomok's love for Mary forms a powerful sentimental force in the novel, as well as a strong figurative link between the characters, for his love, the narrator writes, is "rooted" in his soul. The narrator distinguishes the strength of Hobomok's affection through contrast: "In minds of a light and thoughtless cast, love spreads its thin, fibrous roots upon the surface and withers when laid open to the scorching trials of life; but in souls of sterner mould, it takes a slower and deeper root" (84). Hobomok's strong constitution, in other words, makes him the ideal candidate for Mary's love.

By describing character interiority as a landscape, Child makes affect into a botanical phenomenon. Hobomok's ability to feel love—that ur-force of narrative action—is primed by the soil of his soul. By contrast, as we might recall, Mary discerns the hearts of the Puritan men to be as "harde and sterile as their unploughed soile" (79). Tawil has persuasively argued that Child and other writers of frontier romance helped to forge a new racialized identity based on sentiment, "the notion that members of different races both feel different things, and feel things differently." In this formulation, race "is neither a physiological quality, an intellectual capacity, nor an element of a family history, so much as a psychological and emotional interior."[66] Child renders this interior as terrain and then landscapes it and, in doing so, interpolates land between the body and the sentiment.

In mediating the body through tropes of soil, the narrative creates an internal emotional ecology and grounds feeling in the "mould" of the body. The guiding trope of an emotional landscape within the body renders women powerful cultivators of affective terrain. Mary's response to Hobomok's attention is likewise characterized in these terms: "female penetration knew the plant, though thriving in so wild a soil; and female vanity sinfully indulged its growth" (85). When it comes to affect, feminine intuition penetrates the emotional landscape of the other.

Mary throughout is guided by instinctive reasoning. Looking up at the evening star, she reflects, "Thou hast smiled on distant mosques and temples, and now thou art shedding the same light on the sacrifice heap of the Indian, and the rude dwellings of the Calvinist" (48). The powerful connection Mary feels in this moment is emotional and intuitive, and her proto-Transcendental worldview is founded on recognition of difference and the unity of the natural world. This structure is important for the way Child's novel achieves its message of equality and plurality, for figurative interconnection serves as a leveling characteristic of the American experience. Yet if botanical tropes link all the characters on a rhetorical level, such unity represses troubling historical realities. In

Hobomok we witness the conscious disassociation of national identity from the unpleasant business of material exploitation.

This is perhaps most powerfully expressed in the final line of the novel. After all but erasing Hobomok's presence from the republic, the narrator adds that "the devoted, romantic love of Hobomok was never forgotten by its object; and his faithful services to the 'Yengees' are still remembered with gratitude; though the tender slip which he protected has since become a mighty tree, and the nations of the earth seek refuge beneath its branches" (150). Calling back to Mr. Conant's earlier statement that the Episcopalian scion threatens "our" (Puritan) pleasant plants, this moment reinforces the degree to which the Puritan community had already naturalized ownership over the landscape. The slip here, in other words, allows for a kind of slippage in which the tree's successful cultivation renders it a naturalized authority.

Botanical metaphor thus provides a means of considering Hobomok's role in the nation from the perspective of the horticulturalist. It is a puzzling outlook, for even as he is exiled, his martyrdom makes him a cultivator of sorts, a title with a celebrated status in the young republic. Figuratively imagined not within the tender slip but around it, as a protecting influence, he is much like the frame narrator who is situated not in but around the story being told. The gratitude ascribed to Hobomok at this moment is related to his protection of American identity, a project similarly attributed to the novelist. Consequently, Hobomok's leave-taking at the end of the novel produces ambivalence about the nature of his relationship to American identity. If the nation is merely the tender slip that becomes a mighty tree, then Hobomok is very clearly not incorporated. If, however, we consider the central role of the farmer in early constructions of American identity—Jefferson's agrarianism, for example, or Emerson's embattled farmers—then the final image of Hobomok as cultivator makes him the iconic model of American citizenship.

The botanical imagery up to this point in the novel encourages us to hold two conflicting ideas about Hobomok's positioning in our minds at once: his displacement into a mythical West, tantamount to his erasure from the American scene, and his Lockean cultivation of the soil, which locates him at the heart of the American horticultural project. Reconciling this tension requires reading allegorically and aligning his metaphorical endurance with his material disappearance. Hobomok remains symbolically important to the project of national identity (as constituted through literary narrative) while conveniently dispossessed.

Scholars have elucidated this persistent trope of absent presence in multiple ways. Jodi Byrd has characterized the way in which "Indian-ness" is figured as "transit" in the U.S. colonial and imperial context, "left nowhere and everywhere within the ontological premises through which U.S. empire orients, imagines, and critiques itself."[67] As Richard Slotkin and Alan Trachtenberg have both emphasized, the typecast char-acter of "the Indian" was deeply embedded in white national mytholo-gies across the nineteenth century, and Slotkin stresses in particular how violence between settlers and Native Americans became an animating element of American exceptionalist ideology.[68] In *Hobomok* horticulture serves as a governing metaphor that splits republican ideals from the violence of dispossession.[69]

Child was aware of the ways in which the government was trying to appropriate Native American lands by using farming and cultivation as a marker of ownership. Indeed, four years later she would write a sear-ing criticism of removal efforts in *The First Settlers* (1828), arguing that Cherokees deserved to occupy their land because, according to Lockean logic, they cultivated it.[70] Yet in *Hobomok* Child uses cultivation to jus-tify colonization, naturalizing Hobomok's structural exclusion through his allegorical inclusion.

Child was not alone in devaluing Native sovereignty. As Shirley Sam-uels has pointed out, Jefferson wanted Native Americans to forgo hunt-ing and adapt to Euro-American farming practices because it would reduce the amount of land needed for their sustenance. Consequently, as Samuels interprets Jefferson's logic in relation to his address to the U.S. Congress in 1803, "Forests [for hunting] will thus become available as a form of gift from the tribes who will realize 'the wisdom of exchanging what they can spare and we want, for what we can spare and they want.' The proposal here is built on a condition of translating 'what . . . we want' into a 'wisdom' of unequal exchange."[71] Likewise, in 1823, the year before Child published *Hobomok*, the Supreme Court ruled in *Johnson v. M'Intosh* that Indigenous peoples had no title to the land, building off the earlier logic used by politicians from John Winthrop to James Mon-roe and John Quincy Adams stating that land used for anything other than intensive agriculture did not constitute property.[72] Child explicitly alludes to none of this in the way she closes her novel. By focusing on cultivation as an internal state and linking the landscape to projections of national feeling, she obviates the material debates about land use, ownership, and property that occurred during the first half of the nineteenth century. In this way her novel participates in the ideology of dispossession, for it facilitates the logic by which citizenship is conceived as an emotional property accessed through a proprietary relationship to the landscape.

Hobomok's frame narrative illustrates fidelity to an imperialist horticulture and how thoroughly this ideology patterns the novel. Frederick, the ostensible editor of the text, responds to his friend's desire to write "a New England novel" with incredulity: "'A novel!' quoth I—'when Waverly is galloping over hill and dale, faster and more successful than Alexander's conquering sword? Even American ground is occupied. "The Spy" is lurking in every closet,—the mind is every where [*sic*] supplied with "Pioneers" on the land, and is soon likely to be with "Pilots" by the deep'" (3). The territorial occupation of Scott and Cooper's writing maps the novel onto the landscape, leaving no corner spared: even the sea is vulnerable to annexation, as is the "mind every where supplied" with Cooper's narratives. Rather than seeing this as a threat to his own literary enterprise, Frederick's friend, the ostensibly male anonymous author, turns to horticultural language to claim "his" piece of the literary landscape. By the time that the frame narrative takes place in the early nineteenth century, fields that "blush" into horticultural fecundity valorize the nation on the basis of successful cultivation.

Hobomok thus helps make botanical language central to the body politic and to citizenship. Child's sentimental botanical nationalism here departs from her material understanding of plant circulation as represented in *Evenings*, and her investment in the generic project of a distinctly American historical romance may well account for this difference. Child's use of horticultural symbols in *Hobomok* is part of a familiar allegorical pattern in early American literature.[73] Against fictions that fashion national identity through violence, Child's turn to the botanical offers a pacific alternative mythology of nation building and a relatively open attitude toward racial grafting, even as it continues to rest upon—and legitimate—the logic of dispossession.

Digressive Blooming: *A Romance of the Republic*

Over the four decades between her first and last novels, Child reevaluated botanical nationalism in light of the battles over the freedoms of women, enslaved people, and Native Americans, and she increasingly resisted horticultural practices and rhetoric that fostered exclusion. Drawn toward elements of realism in her writing, and faced with the crisis of the Civil War and its aftermath, Child rejected the symbolic cohesion of many of her earlier fictions by the time she wrote her last novel. Carolyn Karcher has described how *A Romance of the Republic* (1867) "insistently rehearses the history that its white audience was so rapidly forgetting" and tries to articulate a vision of an egalitarian future at a moment when Reconstruction under Andrew Johnson was failing.[74] Most critical approaches to the novel focus on how it advocates racial equality through interracial marriage and a revision of the "tragic mulatta" plot.[75] Recognizing from her own writing and the culture at large how botanical language could service a political agenda, Child returns to plant tropes but this time furthers a more insistently progressive narrative. If cultivation has a nationalist objective in *Evenings* and *Hobomok*, by the time Child wrote *A Romance of the Republic* she was more interested in how to undermine the socially and politically sanctioned racial hierarchy of her first novel's "mighty tree."

The flowers in *A Romance of the Republic* are impossible to ignore. The novel tells the story of Flora and Rosa Royal, daughters of a wealthy white businessman in antebellum New Orleans. The story begins as their father is making preparations to move his daughters to Europe. Unbeknownst to them, they are octoroons, and he is worried about their fate as his health begins to decline. When he dies suddenly, his creditors want to sell the girls as property, but their Italian music teacher, their kind French neighbor, and a younger white man named Gerald Fitzgerald come to their aid and help them escape. Fitzgerald marries Rosa (though the marriage is illegal because he is white and she is technically a slave) and then hides them both in a secret cottage on his island plantation off Georgia. After he starts to make advances toward her sister Flora, Flora flees to the North with the help of an acquaintance. Fitzgerald takes a white woman named Lily Bell as his "legitimate" wife, and both she and Rosa give birth to sons around the same time. The sons are swapped as infants, the sisters both end up in Europe for a portion of the novel (where Rosa becomes a famous opera singer), and nineteen years later they are eventually reunited. At that point their extended families—full of "fresh little flowers in the floral garland" of life—merge.[76] A series of further plot twists involve the denouement of the swapped sibling plot, the death of one half brother, the accidental re-enslavement of the other by his grandfather, and his eventual escape. After the reunion of all living characters when the Civil War ends, the novel closes with a final "German Liederkrantz," or garland of songs, that, after so many plot twists and turns, "closed the ceremonies of the night with Mendelssohn's 'Song of Praise'" (442). Like the ending of *Uncle Tom's Cabin*, the novel's tidy conclusion belies the unresolved challenges of trying to imagine a nation reorganized around the precept of universal rights.

The novel's floral imagery emphasizes the economic motivations behind the desire to enslave human "property." The two daughters are associated with all the educational refinement of Europe before they are rendered Black by virtue of the fact that their mother was a slave. Debra Rosenthal has persuasively illustrated how floral language in the novel functions as racial language, a strategy that engages the historical association of women with flowers in order to generate a floral "counterdiscourse" to the racial rhetoric that connected miscegenation to animality.[77] Rosenthal enumerates the ways that Child capitalized on both scientific and sentimental floral registers in order to align the mixed-race Royal daughters with status-endowing floral discourse associated with white femininity. Part of this entails the deliberate naming of each individual in the novel, whose floral reference—or notable lack thereof—bestows characterological import. Flora, Rosa, Tulee (short for Tulipa), Florimund Blumenthal, Alfred King (a transposition of the King Alfred daffodil), and Lily are only some of the characters framed by a language of flowers with strong metaphorical associations. This framework is undoubtedly helpful in understanding how Child used an available sentimental lexicon to disrupt common scientific constructions of race. At the same time, Child set these floral names against an equally floral tableau, resisting symbolic stability. That is, part of what makes floral rhetoric so powerful in *The Romance of the Republic* is that its ubiquity makes the act of stable meaning making hard to discern. This instability is precisely the point, for the floral imagery in the novel—and it is legion—also serves a metonymic and metacritical function and can be read as references not merely to the codified language of flowers but to the ways in which this language does not fully resolve the range of associations it sets loose.

By 1867 Child was less interested in a fixed, culturally agreed-upon set of floral symbols than in the ways these symbols break down stable epistemologies, including those upon which allegory depends. Very early in the novel the surfeit of botanical imagery overwhelms the narrative and troubles interpretation, as when Mr. Royal invites the younger Mr. King, who is the son of an old friend, to meet his daughters in their New Orleans home. Upon his arrival Mr. King is immediately struck by the beauty of Flora and Rosa, but as they settle into conversation he begins to take in the sitting room as well. The passage is worth quoting at length:

> While the commonplaces of conversation were interchanged, he could not but notice the floral appearance of the room. The ample white lace curtains were surmounted by festoons of artificial roses, caught up by a bird of paradise. On the ceiling was an exquisitely painted garland, from the centre of which hung a tasteful basket of natural flowers, with delicate vine-tresses drooping over its edge. The walls were papered with bright arabesques of flowers, interspersed with birds and butterflies. In one corner a statuette of Flora looked down upon a geranium covered with a profusion of rich blossoms. In the opposite corner, ivy was trained to form a dark background for Canova's "Dancer in Repose," over whose arm was thrown a wreath of interwoven vines and orange-blossoms. On brackets and tables were a variety of natural flowers in vases of Sevres china, whereon the best artists of France had painted flowers in all manner of graceful combinations. The ottomans were embroidered with flowers. Rosabella's white muslin dress was trailed all over with delicately tinted roses, and the lace around the corsage was fastened in front with a mosaic basket of flowers. Floracita's black curls fell over her shoulders mixed with crimson fuchsias, and on each of her little slippers was embroidered a bouquet. (4)

The description runs nearly a page, and perhaps there is no other literary passage quite like it for the kind of saturation in floral decoration that it provides. We are accustomed to long descriptions of interiors in nineteenth-century novels, and taxonomic lists of objects, yet the sheer excesses here untether floral signifiers from any particular artistic, scientific, or sentimental register. In turning a critical eye to the unexamined *stuff* of the Victorian novel, Elaine Freedgood asks what we might learn by paying attention to seemingly banal objects.[78] We might ask a similar question of a tableau so overwhelmingly floral that it prompts Mr. King to marvel, "This is the Temple of Flora. . . . Flowers everywhere! Natural flowers, artificial flowers, painted flowers, embroidered flowers, and human flowers excelling them all" (5).

The surfeit of floral imagery has several effects at once. First, floral decoration here is clearly not simply a domestic enterprise, but an art associated with "the best artists in France" and with Canova, whose statue the daughters have adorned with ivy and orange blossoms. As a form of decoration, flowers here are primarily aesthetic but also scientific, insofar as they invite a kind of perplexed taxonomy. Second, by the time Mr. King makes his comment, the nonhuman flowers have almost threatened to overwhelm the human occupants, and his compliment works by associating the two women with a range of floral typologies. Rosenthal reads this scene as harem-like, especially given the intoxicating incense drifting into the room from the garden outside, adding to the sensory detail. Yet the very diversity of the particulars arrests any attempt to parse how the embroidered flowers on the ottoman differ in meaning and value from the natural flowers in vases, or the painted garland on the ceiling. The sheer variety complicates the association that King makes between the women and the flowers.

It also complicates the idea of flowers as completely natural symbols. Here, natural flowers and floral commodities are entwined just as humans and possessions are at risk of being conflated. The jumbled scene produces a tableau that is hard to organize; there is no centerpiece, and the description takes in each detail of the scene one by one, making their relationality hard to parse. Even Mr. King, who sees the room as a whole, can describe it only in categorical terms that have already been blurred and overrun: natural, artificial, painted, embroidered, human. When everything has a floral association (or is, in fact, a flower), then any attempt at a stable floral order—the kind so present in the moral ordering of *Evenings* and *Hobomok*—becomes meaningless.

The same categorical troubling is true of the plot, whose complexity shows itself to be the product of a disordered society. By the end of the novel, a concatenation of events has occurred that makes discerning heredity so complicated that even the characters must stop at moments and articulate trimmed-down plot summaries to make the connections clear. These events include the swapped babies, siblings who fall in love before their filial relations are revealed, adoption across racial lines, marriages, intergenerational friendships that bestow inheritances, and a whole lot of travel to Europe where names get modified and national identities are confused. The fact that all the favored characters have floral names (including, often, surnames, as in the Blumenthals) and bequeath these names upon children and grandchildren covers the family tree in a profusion of undifferentiated blossoms. When Gerald Fitzgerald Jr. discovers that he was swapped at birth, he comes to see his maternal heritage as double. "I must call you Rose-mother and Lily-mother, I believe," he exclaims upon learning the news (364). His recognition of both women as his mothers suggests that maternity—and by extension identity, race, and citizenship—cannot be reduced to biology any more than Native Americans can be reduced to allegorical figures who acquiesce in land appropriation.

The "Temple of Flora" that occupied so much narrative description early on becomes a reference point for the way that self is conceived in this narrative. When Rosa and Flora first learn that they must leave home after their father's abrupt death, the space is described as constitutive of their identities: "The garden and flowery parlor, which their mother had created and their father had so dearly loved, seemed almost as much a portion of them as their own persons" (41). This merging of character with these spaces proves an important point of reference as the novel proceeds. Shortly before Rosa and Alfred King become a couple, she asks him if he remembers the "flowery parlor" where he first met her and her sister. His response emphasizes the room's role in mediating his memory and desire: "I see that room as distinctly as you can see it. . . . It has often been in my dreams, and the changing events of my life have never banished it from my memory for a single day" (250). The fact that the room resurfaces during this key plot turn speaks to its importance in the way that the floral references shape the narrative arc. Flora and Rosa's strong identification with the parlor ties them to a multitude of botanical representations but to no singular floral custom.

In *A Romance of the Republic*, Child recognizes the limits of her earlier botanical imagination and draws attention to how horticultural language could serve prejudicial ends. At the end of the novel a Mr. Bright makes an observation about his distaste for Boston society by critiquing its landscape management practices: "I don't generally like to go among Boston folks," he says. "Just look at the trees on the Commons. They're dying because they've rolled the surface of the ground so smooth. That's just the way in Boston, I reckon. They take so much pains to make the surface smooth, that it kills the roots o'things. But when I come here, or go to Mrs. Blumenthal's, I feel as if the roots o'things wa'n't killed" (441). For Bright, and for the narrative position he represents, a homogenous landscape is a dead landscape. By contrast, the two sisters' homes sustain something vital.

Bright's longing for the "roots o'things" sets up a social structure of surface versus depth, but it also reveals the extent to which botanical tropes come to govern social expression, shaping ideas of self and society. When Alfred King comments early on that "when men get to be so old as I am, the process of being transplanted in foreign soil seems onerous" (18), he relies upon a botanical process that would have been familiar to readers. *A Romance of the Republic* reveals how organic metaphors could not only constitute a world but point to its disorder. For the reader conditioned to interpret botanical imagery taxonomically, this proliferation of value is disruptive. Rather than constituting a holistic botanical system, as in her early works, the varied kinds of botanical representation here disrupt a stable moral or aesthetic order. Unmooring a literary correlation between botany and human biology familiar to her readers, Child's last novel seeks a new political order that resists smoothing over the complex roots of things, deadening future growth.

* * *

In understanding horticultural practices as a strategy for national identity construction, Child's early work plots a connection between nationalism and cultivation that is material and metaphorical. In *Evenings* this takes the form of a meticulous catalogue of imported plants and an emphasis on the importance of economic botany to domestic comfort. In *Hobomok* it manifests as botanical allegory that turns horticulture into the basis of a nationalist—and in the final assessment imperialist—sentiment. Child's early work establishes horticultural language as foundational to civic identity and is largely uncritical of the way that acts of cultivation shored up settler colonialism. While she eventually uses botanical language in *Evenings* to critique the slaveholding society in which she lives, the novel reifies the nation as its horizon of care.

Child's botanical depictions of America were part of a broader societal effort to understand the polity in natural terms. In literary spheres nationalism has largely been analyzed in relation to print culture, given the ways that print technology shaped connection and imagined communities across distances and regional divides.[79] Print culture also played a role in threatening a unified idea of the nation, as Trish Loughran has compellingly argued.[80] The materiality of print in this period was integral to the ways individuals conceptualized larger networks and their own connectivity within the republic. But so too was the materiality of the landscape; authors and readers alike turned to acts of transplantation and cultivation to imagine national sentiment. Critics have long described how migrant farmers located in western territories understood their activities in national terms and how organizing metaphors of American wilderness promulgated the idea of an exceptional environment. But the idea of exceptional wild nature was not the only one fueling a sense of national pride.[81] The correlation of environment with nation also played out in the domestic sphere. Pedagogical materials for children articulated the political economy of trees. Botany manuals sought to compare and contrast native flora from European botanicals. And writings explicitly correlated American sentiments with flowers that could grow in garden or greenhouse.

The chapters that follow expand on the ways that botanical nationalism flexed to accommodate rapid developments in plant science and technology. As the next chapter shows, horticultural technology increasingly allowed for tighter control over botanical life in gardens and greenhouses. Such regulation was celebrated both for its role in shaping the nation's political economy and as a model for ideas of (self-) mastery espoused by midcentury moralists and aligned to the nation's biopolitical regime. But plant movement could never be fully regulated, a fact that home gardeners understood through experience, and that Nathaniel Hawthorne, experienced with customs houses and horticultural clubs, translated into literary narratives of disruption.

2

Botanical Disruption

In 1840 a grape seedling growing near a wall on a Concord property caught the eye of proprietor Ephraim Bull. A gardening enthusiast, Bull transferred the seedling to his garden in order to give it "good cultivation," as one article later described it.[1] The seedling took in the new soil and after several years the vine was fertilized—perhaps by a nearby Catawba variety he had planted—and produced several bunches of fruit. American horticulturalists had experienced many difficulties trying to successfully cultivate Europe's popular *vinifera* species, and Bull was committed to hybridizing grapes from New World vines that could compete with *vinifera*'s flavor, versatility, and emblematic status. He saved the seed from this hybrid and launched a multiyear project to produce a sturdy grape with a sweet reputation. Along the way there were countless failures. Then finally, in the early 1850s, Bull worked with the nurseryman Charles Hovey to bring a satisfying new variety to market. He named it the Concord, connecting the grape to his town, its landscape, and its people.

In the 1840s Concord was full of horticulturalists like Bull who sought to "improve" natural varieties of fruit by hybridizing for desired qualities. The founding of the Massachusetts Horticultural Society in 1829 had increased the distribution of seeds and conversation about best cultivation practices. Gardeners traded seeds with friends and colleagues and had ready access to nurseries and horticultural clubs that could outfit them with the latest varietals from the United States and Europe. Horticulture became increasingly organized during the second quarter of the nineteenth century, with efforts to successfully grow fruit tied to a belief that these efforts would attest to the successful development of American culture.[2] While the U.S. government did not officially begin to regulate the introduction of new species until the 1890s, horticultural institutions sought to import only plants they deemed would be a credit to the country, and individuals hybridized with similar goals in mind.

Experimenting with plant life became something of a Concord pastime. Even before he named the new variety after the town, Bull shared some of the seeds with his neighbors there. One of them was Nathaniel Hawthorne.[3]

As an amateur gardener, Hawthorne witnessed many major horticultural changes across his lifetime, and his writing reflected one of the most prominent features of these efforts at domestic cultivation: failure. While newspapers and horticultural halls were full of prize varieties and stories of success, gardeners were familiar with the fact that for every Concord grape there were countless forgotten fruits. Growers often could not understand why one seed sprouted and another did not, or they might care for a plant only to watch it wither on the vine. In the notebook she shared with her husband, Sophia Hawthorne described her excitement at "the first time I ever put any seeds into the earth" in May 1843. Six weeks later she was reflecting on the mixed results: "My garden does not entirely succeed. Not all the seeds I planted have come up & but one dahlia of the five which Father put in the ground when he was here the second week of May."[4] Sophia quickly moved on. The challenges that plants posed to human gardening efforts tended to be overshadowed by the successes. But for her husband, as for many who tended the soil, the theme of botanical difficulty persisted like a stubborn weed. Hawthorne's fiction and personal writings provide a window into a culture anxious about the meaning of plants that failed to behave as anticipated.

Such anxiety was nothing new, of course. Jefferson had worried about the issue in the previous century, vexed about what agrarian failures might mean for the new republic. But in Hawthorne's era such worry was coupled with means of regulating plant life—horticultural societies, soil amendments, and glass technologies—that were increasingly accessible to the middle class. Observing the growth of plants in these contexts, gardeners experienced how plants nonetheless continued to thwart expectations. Even good culture did not ensure success or expected outcomes. Plants that behaved in surprising ways or failed to thrive disrupted popular conceptions of plants as stable symbols of societal values. Educators, moral reformers, and writers frequently invoked cultivation as a master metaphor for establishing appropriate behavior for women and men.[5] But this assumed plants obeyed human designs.

Across his career, Hawthorne traced the frustrations and possibilities that his contemporaries saw in disruptive plant behaviors, examining how authority responded to plants' own unpredictable strategies.

Human-Plant Relations in Antebellum America

In his 1879 biography, Henry James characterized Nathaniel Hawthorne as a quintessentially regional writer: "Out of the soil of New England [Hawthorne] sprang—in a crevice of that immitigable granite he sprouted and blossomed. . . . Hawthorne's work savours thoroughly of the local soil—it is redolent of the social system in which he had his being. This could hardly fail to be the case, when the man himself was so deeply rooted in the soil. Hawthorne sprang from the primitive New England stock; he had a very definite and conspicuous pedigree."[6] To take James's view, which became a bellwether for subsequent interpretations, Hawthorne was as New England as the soil itself.[7] If Hawthorne's themes register on the level of national politics, they are nonetheless always checked by a nature that is, at heart, "exquisitely and consistently provincial."[8]

In characterizing Hawthorne as a hardy New England flower, James makes the point that the author's worldview is rooted in one spot. Provincialism in James's formulation is akin to soil and stock—biological matter that is synonymous with place. Plants would seem to furnish the imagination with ready metaphors for attachment and entrenchment. After all, they grow roots; to plant something is to stop it from moving. But for all that James invoked plants in this manner, Hawthorne was one of many gardeners whose actual experience of plants and soil starkly contradicted provincialism. Rather than fixtures, plants in antebellum America were often understood as part of new channels directing the flow of information, goods, and people around the world. And in consequence, they were a primary way that Americans understood new geopolitical realities, trade networks, and material, human, and botanical diasporas. The irony of James's formulation is how it misses the ways in which Hawthorne himself used plant tropes to engage with the mobility and flux of nineteenth-century plants.[9]

Hawthorne came of age in a time when botanical commodities and scientific specimens from around the world frequently arrived in the port cities where he spent most of his life. The biotic composition of New

England changed substantially across the eighteenth and nineteenth centuries, and Hawthorne himself witnessed a dramatic transformation of plant practices during his lifetime as botany became more industrialized and institutionalized. One of his uncles, Robert Manning, was a prominent pomologist who participated in the founding of the Massachusetts Horticultural Society, and Hawthorne gained early exposure to gardening and arboriculture through him.[10] Manning introduced many new varieties of fruit trees from around the globe to family residences in Salem, Massachusetts, and Raymond, Maine, where Hawthorne spent some of his happiest times as a child. And he roomed in Boston for a short while with Thomas Green Fessenden, a farmer, statesman, and editor of several horticultural periodicals.

Manning's focus on introducing novel plant varieties was part of a broad effort to improve public taste. In 1835 the first president of the Massachusetts Horticultural Society described the society's significant role in obtaining the best trees, cuttings, and seeds from Europe, Asia, South America, and the United States and singled out Manning's efforts in this regard.[11] Newly imported plants were increasingly the subject of horticultural display aimed at improving taste and morality. A June 1855 *New York Times* article called horticultural exhibitions "the most profitable [contrivances] and the least capable of being perverted to an improper end," noting how they functioned "to exert a healthful influence on public taste and moral."[12] Such displays demonstrated the viability of the soil and the intelligence of its cultivators, reifying the notion that the landscape, when approached with scientific skill, was fully manageable.

The idea of America as a machine-made pastoral has been the subject of much scholarship since Leo Marx's 1964 classic *The Machine in the Garden*. Yet for Marx, as for more recent critics like Richard Slotkin, the primary focus is on technological impositions *onto* the landscape in the form of railroads, factories, smokestacks, and the like.[13] Hawthorne's depictions of the organic realm, however, reflect how technology was transforming the very stuff of nature itself. More pointedly, during the 1840s and 1850s, when Hawthorne was most prolific, English landscape gardening, which strongly influenced antebellum garden design, put the increasingly complex relationship between plants and industrial technology on display.

The trope of the English technological garden can be traced back at least two centuries to Andrew Marvell's "The Mower against Gardens," where the gardener "does nature vex, / To procreate without a sex" and the botanist must ensure the survival of the *"Marvel of Peru"* when transplanted in English soil.[14] By the Victorian era the gardener's growing power to vex nature—and prune it, force it, train it to a wall—increasingly relied upon the same material technologies associated with urban development: pipes, engines, and cheap glass. The invention of an iron hinged sash bar, for instance, by Scottish horticulturalist John Claudius Loudon, facilitated greenhouse construction and made possible both the winter cultivation of temperate plants and the importation of plants from warmer climates.[15] Technology was not so much invading the garden as constituting it in the first place.

Such dramatic transformations in plant culture were not lost on Hawthorne, whose literary interests in the careful optics and management of gardens is evident from his early short story "Rappaccini's Daughter" (1844) through his later depictions of aberrant plants that flaunt the human desires so meticulously invested in them. The dynamics of the Padua garden that Hawthorne describes in "Rappaccini's Daughter"—the careful sight lines and strict limitations on movement and touch—reflect the painstaking construction of nature as a scripted experience. Much like Poe's later "The Domain of Arnheim" (1846) or the palm stove at Kew, Hawthorne's story describes an environment that is contrived down to its last detail.[16]

Hawthorne criticism has often portrayed him at a remove from nature, interested in it only to the extent that it served his allegorical purposes. Perhaps this is because Hawthorne often complained—in his journals, letters, and *The Blithedale Romance* (1852)—that the work during his short stint at Brook Farm left him little time for his writing. In fact, Hawthorne was an avid gardener, appreciating the tangible rewards of his horticultural efforts in Concord. In a notebook entry dated August 10, 1842, Hawthorne provides a brief account of the pleasures of growing vegetables at the Old Manse, "as if something were being created under my own inspection, and partly by my own aid." He further notes, "I find that I have not given a very complete account of our garden; although, certainly, it deserves an ample record in this chronicle; since my labors in it are the only present labor of my life."[17] Hawthorne's notebooks are full of scattered and relatively inconsistent reflections on his garden and the wild and cultivated landscapes he encounters around

Salem and Concord. In an outburst that sounds almost Thoreauvian, he writes on September 4, 1842, "Oh that I could run wild!—that is, that I could put myself into a true relation with nature, and be on friendly terms with all congenial elements" (358). More often than not, however, Hawthorne reflects on the process of cultivating land and describes an intimate relationship between the landscape and the cultivator. Shortly after inheriting an apple orchard at the Wayside in Concord, Hawthorne writes in his diary on August 9, 1852,

> My fancy has always found something particularly interesting in an orchard—especially an old orchard. Apple-trees, and all fruit-trees, have a domestic character, which brings them into relationship with man; they have lost, in a great measure, the wild nature of the forest-tree, and have grown humanized, by receiving the care of man, and by contributing to his wants. They have become a part of the family; and their individual characters are as well understood and appreciated as those of the human members. One tree is harsh and crabbed—another mild—one is churlish and illiberal—another exhausts itself with its free-hearted bounties. (327)

Hawthorne saw plants, once domesticated, as part of the family, that most intimate and exacting of nineteenth-century social institutions.

Hawthorne was not alone in understanding horticulture in familial and domestic terms. Evidence of the intimate relations between domestic cultivators and their plants is abundant in periodical culture from the mid-nineteenth century. For Hawthorne's indomitable, mercurial friend Margaret Fuller, home itself was marked by the transplantation of familiar plants to new locales. In *Summer on the Lakes* she is moved by "families we saw [who] had brought with them and planted the locust." Describing it as "pleasant to see their old home loves brought in connection with their new splendors," she connects the "tenderness of feeling" revealed by this act to "prosperity and intelligence, as if the ordering mind of man had some idea of home beyond a mere shelter, beneath which to eat and sleep."[18]

For Hawthorne's contemporary and competitor Stowe, plants in the home "are a corrective of the impurities of the atmosphere," and she cautions that "it is a fatal augury for a room that plants cannot be made to thrive in it. Plants should not turn pale, be long-jointed, long-leaved, and spindling; and where they grow in this way, we may be certain that there is a want of vitality for human beings."[19] Hawthorne engaged this same theme in "The Birthmark" (1843) when Alymer tests his fatal concoction on a plant before giving it to Georgiana. Both authors addressed plants not simply as beneficial to humans but as material proxies for human health.

The relationship between plant culture and home life in antebellum America was likewise evident in the proliferation of botanical content in popular weeklies, as well as the rise of a number of magazines and journals devoted to gardening and horticulture. As one article in an 1837 volume of the *Farmer's Cabinet* suggested, "The pleasure to be derived in cultivating flowers can now be appreciated by most persons, as their biography and science have become household ornaments."[20] For middle-class readers, plants and the family were increasingly connected in sentimental culture, a situation that Hawthorne explored most directly in his early short story "Rappaccini's Daughter."

"Rappaccini's Daughter" and Domestic Science

Though moralists largely praised ornamenting a home with houseplants, the changing nature of the domestic garden also worried some, especially as cultivation increasingly involved scientific intervention.[21] "Rappaccini's Daughter" foregrounds a number of issues that horticulturalists were actively navigating, including the impact of botanical science on family dynamics. Literalizing the kinship between humans and plants, Hawthorne anxiously narrates how horticultural technologies were shifting the meaning of home gardening practices. The story presents a vision of how new technology might unsettle gendered domestic relations and patriarchal authority. Horticultural engineering promised control over nature, but it also held the potential to destabilize gendered floral symbolism. The story's garden, "cultivated with exceeding care," is a space of both domestic affiliations and panoptic spectacle that, upon examination, produces fascination and horror.[22]

These emotions largely belong to Giovanni, a young scholar who arrives in Padua and falls in love with a woman named Beatrice who tends to the private garden below his window. Her father, a renowned botanist named Rappaccini, has raised her to tend to the poisonous plants in the garden, and she has consequently become physically imbued with their toxic qualities. Giovanni's satisfaction in assuming that "this spot of lovely and luxuriant vegetation . . . would serve . . . as a symbolic language, to keep him in communion with Nature" (393) cedes to terror when Nature in the garden does not align with his expectations. When Giovanni accuses Beatrice of rendering a poisonous influence on him, she swallows a rival botanist's antidote and dies, so central has the poison become to her own vitality. The short story is an allegory for the dangers of cold empiricism and scientific megalomania as well as for racial "commixture," as Anna Brickhouse has pointed out.[23] But it also registers the fact that institutional botany and domestic gardening were increasingly entangled enterprises, unsettling Giovanni's idealized vision of Beatrice and rendering her inscrutable.

Technologies for plant circulation facilitated intimate domestic relations with plants from all over the globe, and these domestic gardening projects increasingly destabilized the gender politics of space and science.[24] Prior to the early nineteenth century, scientific gardens, where the male empiricist was sovereign, were perceived as distinct from domestic gardens where women might reign supreme. Botanical gardens tied to universities or governments, and the physic gardens that had since the sixteenth century preceded them in Europe, were generally perceived as masculine spaces, repositories for medical experimentation and eventually also for the spoils of colonial bioprospecting. The "domesticated virtuous garden," in contrast, was perceived as a site of education that offered a healthful, safe place for the cultivation of feminine virtue and moral development for children.[25] "Rappaccini's Daughter" conflates these functions, demonstrating how the domestic garden was increasingly a site of botanical innovation that collapsed the gendered garden binary. After reasoning that the garden he gazed down upon was an institutional botanical garden, Giovanni adds, "Or, not improbably, it might once have been the pleasure-place of an opulent family" (389). In any case, it is now part of the home of Beatrice and her father Rappaccini. Here the familial and the institutional overlap just as the empirical and sentimental converge.

As critics have argued, Beatrice's kinship with the "monstrous" plants threatens to corrupt the idea of bodily purity. Because Beatrice has such a close correspondence with the flowers, her body becomes the site of anxiety about racial mixing. Brickhouse shows how Hawthorne emphasizes the hybridity of the plants to raise the specter that their sister Beatrice is similarly "the product of a kind of botanical miscegenation."[26] Dana Medoro similarly reads Beatrice's alignment with crossed plants as an anxious symbol of white purity that must be maintained at all costs.[27] In both cases hybrid plants are fearful, in sharp contrast to the positive analogies that figures like Child and Stowe drew between botanical hybridity and race mixing.[28]

If racial purity is in the short story's foreground, the colonial history of plant circulation represents its crucial backdrop. Padua was the site of one of the world's first botanical gardens, a fact that features in Giovanni's first assessment: "From its appearance, he judged it to be one of those botanic gardens, which were of earlier date in Padua than elsewhere in Italy, or the world" (389). The purpose of Padua's botanical garden, founded in 1545 as a horto medicinale, was to serve as a repository of therapeutic plants. In setting the story in Padua, Hawthorne not only displaces it from the Americas but situates the story in a place significant precisely for its botanical garden. By the nineteenth century the Padua garden was known as the first of many such institutional gardens, Kew being only the most celebrated. In this light, the garden is tied to a lineage of colonial horticulture, characterized by the collection and mixing of plants from around the globe. The original Padua botanical garden, in fact, was circular, shaped to represent a globe. Placing the story in Padua emphasizes the potency of botany as a tool of global control.

While critics have repeatedly stressed that the domestic sphere was idealized as a stabilizing context for the individual, the enclosed garden in "Rappaccini's Daughter" is a precarious site. The garden's experimental plants assert that human engineering is constitutive to its very existence. Deirdre Lynch has described the ways in which climate control and greenhouse technology began to shape ideas of nature in the Romantic period. She further illustrates the effect of this "*greenhouse romanticism*" in unsettling nature's role in authenticated courtship and marriage.[29] In overlaying the affective relationship between Beatrice and plants with the botanical discourse about plant research, Hawthorne links sentimental culture's association between women and plants with the latter's scientific circulation.

By 1807 greenhouse plants were available in Salem, and by the 1830s and 1840s, as Hawthorne's cousin Robert Manning Jr. describes, steamships had transformed horticulture in both the United States and Europe "by the opportunity which it afforded for the interchange and concentration of the fruits and flowers of every climate."[30] As attention shifted to these plants, so too did interest in the measures taken to keep them alive in a new climate. One amateur correspondent writing to William Hooker, the director of Kew, noted that she had seeds from an interesting pine that she would happily send him, and "in return would be very grateful for any hints that Sir William would kindly give as to the profitability? And the means of creating a soil in this country that would suit them."[31] Cultivation connected the professional class of botanists at institutional gardens like Kew with the amateur gardeners around the globe.

The construction of large-scale greenhouses in European botanical gardens starting in the 1820s relied upon artificial methods like boilers and elaborate piping systems in order to sustain tropical plants in colder climates. Freiburg's botanical garden developed an impressive greenhouse in the late 1820s, and Charles Rohault de Fleury's design for two greenhouses at the Jardin des Plants in the mid-1830s was based on careful study of greenhouse construction in England. Shortly after Kew became a publicly financed institution in 1841, plans began for the Palm House, a "stove" (or heated greenhouse) for the cultivation of palms that required careful temperature regulation. One characteristic newspaper account of the new building notes,

Heat is communicated by means of hot-water pipes, extended to a length of 24,000 feet, and concealed from view under the tables and beneath the floor. The furnaces, twelve in number, are underground, and the smoke is conveyed from them through a subterranean tunnel, to a distance of 479 feet from the house, where it escapes by a shaft of ornamental structure, 96 feet high, with a reservoir near the top for supplying the house with water. The coals and ashes are carried to and from the furnaces through the tunnel; so that everything that could offend the eye is carefully concealed, and the plants are protected from the injurious effect of dust and fuliginous matter. The useful and the ornamental are combined with exquisite taste.[32]

The extensive efforts to conceal the offending instruments used to heat the greenhouse and the equally extensive descriptions of these efforts in print culture together primed readers to consider the relationship between industry and horticulture as a marvelous, tasteful, and above all natural union. The concealed tunnels running under the grounds at Kew spoke to an interest not only in regulating climate and controlling nature but in using technology to service botanical spectacle. Yet importantly, the technological aspects that were so carefully hidden out of sight—the tunnel for the transport of coal, the smoke from the boilers and everything else "that could offend the eye"—were the subject of much fascination in periodical culture. People wanted to know how the greenhouses worked from an engineering perspective and to understand the technological apparatus that made possible the display of horticultural grandeur.

Published at a moment when botanical gardens were becoming public institutions, "Rappaccini's Daughter" dwells on these themes of human engineering and horticultural display. The plants in the garden are spectacular and spectacularly poisonous. In contrast to the positive valence of horticultural hybridity in botanical magazines and seed catalogues, Hawthorne characterizes the "adultery of various vegetable species" as "the monstrous offspring of man's depraved fancy, glowing with only an evil mockery of beauty" (404). Rappaccini dons the "armor" of gloves and a face mask while pruning "flowers gorgeously magnificent," and even then suffers in proximity, summoning his daughter Beatrice to tend to the flowers (391, 389). Beatrice, by contrast, is inured to the vegetable toxins but toxic herself, giving Giovanni the impression that she is likewise "to be touched only with a glove, nor to be approached without a mask."

Beatrice's kinship with the plants is shown to be threatening both because it makes her "poisonous" and because it illustrates her own scientific expertise. In many respects, Beatrice's "bloom so deep and vivid" is part of a long narrative tradition that uses the rhetoric of bloom to describe women on the cusp of sexual maturity, but in the botanical garden this makes her "fraught with some strange peril."[33] Beatrice's sisterhood with the plants growing in the garden is certainly an analogy for miscegenation, but also a way that Hawthorne demonstrates Beatrice as a sympathetic cultivator. This kinship to the plants in the garden has received more critical attention than her scientific acumen. Beatrice's intimate relationship to the garden is coupled with her empirical practices, making her, the story suggests, a better scientist than her father.

The botanical garden in the story is still a domestic area, where Beatrice appears like "another flower, the human sister of those vegetable ones" (391) in a space where "every portion of the soil was peopled with plants and herbs" (389). Yet it is also a site of "fatal science" where many of the plants that grow have "an appearance of artificialness" as "the result of experiment." In contrast to Beatrice, Rappaccini allows "no approach to intimacy between himself and those vegetable existences" (390). Hawthorne stresses that this affective difference between father and daughter hinges upon the former's "look as deep as Nature itself, but without Nature's warmth of love" (401). It would be easy to declare that Hawthorne is merely telling an allegory of the force of science—personified in the great botanist—over the force of nature, if not for the way that "Rappaccini's Daughter" destabilizes this oppositional dualism by rendering science and nature deeply intertwined, and by making Beatrice the central figure for negotiating that interaction.

"Rappaccini's Daughter" diagnoses the inability to confine scientific experimentation to the sphere of male professionals. Beatrice, for all she urges Giovanni to "not believe these stories about my science," is an empiricist, urging him to "believe nothing of me save what you see with your own eyes" (405). Yet what Giovanni sees with his own eyes he fears "deceive[s] him" (398), as when the flowers he tosses down to Beatrice wilt or the insect buzzing round her head falls to the ground. In shaping the garden she does not passively follow the directives of her father, but as rival botanist Baglioni points out, "she is already qualified to fill a professor's chair." He goes on to add, "Perchance her father destines her for mine!" speaking to a real anxiety about her ability to outperform him in the academy (395). Hawthorne describes a woman without a love and taste for flowers as a "monster," but the men in "Rappaccini's Daughter" see something equally monstrous in Beatrice's intimate relationship to the plants when cultivation is framed as a matter of plant science. When plants appear "fierce, passionate, and even unnatural," they cease to be the unthreatening signifier of women's sentimental nature (403).

Beatrice's superlative ability to apprehend nature in the garden is part of the story's preoccupation with gendered access to natural knowledge.

Her sisterhood with the plants is not merely physical but affective as well; her identification with the plants is familial and rooted in care. And while Beatrice distinguishes between her body "nourished with poison" and her spirit as "God's creature" (418), the story represents her contagious effect on Giovanni as the result of science mixed with sympathy. In other words, Rappaccini's influence is physiological and affective. When Giovanni recoils at Beatrice's "poisonous" influence on him, both father and daughter identify the power of that poison as "sympathy." Toward the end of the story Beatrice exclaims to Giovanni by way of explanation "my father!—he has united us in this fearful sympathy!" (418). A short while later Rappaccini likewise explains to his daughter, "My science, and the sympathy between thee and him, have so wrought within his system, that he now stands apart from common men, as thou dost . . . from ordinary women" (419). Sympathy conveyed a structure of meanings in nineteenth-century America with incredible political and affective power.[34] As Glenn Hendler notes, a sympathetic attachment between subjects invites intimate identification, a mediation "between a distanced observer and the sufferer [that] is always at risk of collapsing."[35] But here sympathy is rendered as a kind of contagion or infection—something that operates materially and has material consequences on Giovanni's body. Permeating the body, it collapses the idea of the discrete, individuated subject.[36]

This is terrifying to Giovanni, a terror that is compounded as he acknowledges his similarities with other forms of life. While the garden is walled off from the city, it is not hermetically sealed. Human entrance and egress are carefully controlled—Rappaccini and Beatrice enter the garden through their home, and Giovanni learns about a hidden second entrance only from his landlady—but the garden walls do not prevent insects and reptiles from entering. All of Giovanni's fears about Beatrice's poisonous nature are confirmed when he watches insects, reptiles ,and flowers from outside the garden come in contact with the atmosphere within. First is a "small orange-colored reptile, of the lizard or chameleon species" that, in creeping along the garden path, comes in contact with a drop of moisture from the broken stem of one of the poisonous flowers and subsequently "contort[s] itself violently," and dies (397). Subsequently "there came a beautiful insect over the garden wall; it had perhaps wandered through the city and found no flowers nor verdure among those antique haunts of men, until the heavy perfumes of Doctor Rappaccini's shrubs had lured it from afar" (397). Giovanni watches as it circles Beatrice and then faints dead "from no cause that he could discern, unless it were the atmosphere of her breath" (398). Next Giovanni throws a bouquet of flowers down to Beatrice, and from a distance discerns that they begin to "wither in her grasp" (398). These successive scenes of exposure confirm for Giovanni the toxicity of both the plants and Beatrice.

More fundamentally, they also confirm a shared vulnerability across species. Fearing he has been contaminated, Giovanni breathes on a spider in his chamber and finds that it "suddenly ceased its toil." After another breath it "made a convulsive gripe with his limbs, and hung dead across the window." The spider serves as proxy for his own condition. Put another way, Giovanni comes to anticipate his own poisonous nature through the consequences that play out on other life forms. The story spectacularizes their shared vitality and treats the idea itself as poisonous. At the end of the story three men watch as Beatrice, suspecting it may kill her, chooses to drink the antidote provided by Giovanni. Beatrice's kinship with plants extends beyond their poisonous nature to their capabilities of defiance, and it is in this light that Rappaccini's scientific success is rendered as a domestic failure.

Weeding in the Garden of Good and Evil: *The Scarlet Letter*

The kinds of engineering displayed in "Rappaccini's Daughter" produced fears about intervening too much in the plant realm, and especially the power women were capable of wielding in this manner. At the same time, as the story shows, horticulturalists conversely harbored fears that they would not be able to control the plants under their care. Cultivation has a long history as a trope for socialization and proper behavior, but gardeners regularly confronted plants that failed to realize their expectations. Plants that scatter and circulate threaten the stability of organic metaphors for civil obedience, circumstances strikingly apparent in *The Scarlet Letter* (1850), a novel that is often read as an object lesson in consent. As Laura Korobkin has described Hawthorne's Puritan New England, "The people may mutter, but they must also unhesitatingly obey."[37] According to critics such as Sacvan Bercovitch and Lauren

Berlant, Hester's consent to the punishment imposed by the collective becomes the exemplar for an American individualism that promotes cohesion by outward acquiescence.[38] But in terms of the novel's botanical imagery, Hawthorne takes the domestic cultivation narrative and destabilizes it by highlighting the ways that plants act in unpredictable, autonomous ways.

The Scarlet Letter resists the didactic moral symbolism prevalent in popular plant culture while also playing on plants' radical potential to inspire aberrant feeling. At a famous moment in "The Custom House" preface to the novel, the narrator declares that "human nature will not flourish, any more than a potato, if it be planted and replanted, for too long a series of generations, in the same worn-out soil. My children have had other birthplaces, and, so far as their fortunes may be within my control, shall strike their roots into unaccustomed earth."[39] The notion of unaccustomed earth proves relatively elusive in a novel where the mark of Hester's adultery is portable property that she wears, where Dimmesdale refuses to leave the settlement, and where Pearl ends up on European soil, which Hawthorne certainly would not have considered "unaccustomed." This passage, however, suggests the degree to which plants provide Hawthorne with a model for domestic relations that are organic but not rooted, allowing him to reinterpret the plant tropes commonly used to model socialization. Whereas conventional botanical models of behavior stressed tropes of careful tending, pruning, and training, Hawthorne shows how plant growth that resists these efforts modeled other possibilities. In this sense, his botanical images run counter to the message of conformity so central to both the way the novel is often read and the way horticulture frequently operated as a tool of educators and theologians tasked with socializing the young.

Plants were a favorite disciplinary metaphor among antebellum moralists and educators. For instance, in one of his *Lectures on Education* (1845), the educational reformer Horace Mann described collective responsibility for education thus: "And this amazing change [wrought by education] in these feeble and helpless creatures,—this transfiguration of them for good or for evil,—is wrought by laws of organization and of increase, as certain in their operation, and as infallible in their results, as those by which the skillful gardener substitutes flowers, and delicious fruits, and healing herbs, for briars and thorns and poisonous plants. And as we hold the gardener responsible for the productions of his garden, so is the community responsible for the general character and conduct of its children."⁴⁰ The passage offers a clear link between cultivation and morality. Such analogies between gardening and child-rearing were plentiful in popular periodicals from the 1830s and 1840s and were directed especially at the young. Encouraging children to understand their actions in relation to a horticultural framework, an article in the May 1838 edition of the *Youth's Companion* urges: "Children, when you feel inclined to be selfish, when you are angry, or when you are fretful and sullen because you have been forbidden something you wished to do, and when you feel envious of another and covet something which he has, then there has a time come when you may choose whether you will be like a bad and *poisonous plant*, which everybody avoids and wishes away, or a good and wholesome plant, which we cultivate carefully, and love to have near us."⁴¹ This choice is no choice, and the tone here perfectly encapsulates the disciplinary intimacy that Richard Brodhead identifies as operating within antebellum domestic and educational institutions.⁴² The stakes of proper cultivation were frequently pitched as nothing short of the future of the nation. An article titled "On Education" written for the *Cultivator* in 1836 makes an explicit comparison between a nation's "two natural sources of wealth: one, the *soil* of the nation, and the other, the *mind* of the nation."⁴³ In the process of moralizing on the distinction between plants and weeds, the article—and even the very title of the publication—aligns mental cultivation with the act of tending the soil, framing the proper cultivation of both mind and land as essential to national prosperity.

Nineteenth-century readers also had no shortage of fictional plants that helped teach individuals how to fall in line. One of the most widely read novels in the mid-nineteenth century was *Picciola*, by French writer X. B. Saintine (the nom de plume of the French writer Joseph-Xavier Boniface), about a man who falls passionately in love with a flowering plant. Published in 1836 and translated into English in 1838, the novel became a best seller in America almost overnight. An 1839 review in *Ladies' Companion* calls *Picciola* "the most striking and original tale that has appeared in our country for a long time" and notes that the book passed through four editions in its first month of publication.⁴⁴ By 1847, reviews in a wide range of periodicals—from women's magazines to agricultural journals—were already hailing it as a classic.

The novel that reviewers praised with such enthusiasm offers a redemption plot set in the immediate aftermath of the French Revolution. Count Charles Veramont de Charney is a wayward French aristocrat who takes aim at the state, goes to prison, and there falls in love with a flower that grows up through the stones of the floor. His journey to freedom is a matter of reformation caused by this vegetable love, an act that politically neutralizes him in the eyes of the French state. The emperor eventually pardons Charney by rationalizing that "he who could submit his powers of mind to the influence of a sorry weed, may have in him the makings of an excellent botanist, but not of a conspirator." Indeed, Charney's study of nature—concentrated on this one particular plant—nullifies his radicalism. Charney himself reflects on the unimportance of Napoleon's newest conquests in comparison with an insect that threatens his beloved plant. Charney's avocation both neutralizes his threat and also suggests his potential worth to an empire that prized botanical spoils. As one 1837 review noted, in pardoning a repentant Charney, "Napoleon secures a botanist the more for France, instead of a plotter against the tranquility of the state."[45]

Charney's eventual freedom corresponds with marriage to the daughter of another prisoner, and the final chapter finds Charney happily ensconced in a set of bourgeois domestic relationships. The plant thus ultimately becomes a means to an end as an object lesson in domestic values. The conservative conclusion offers a lesson about how plants might foster family feeling, and the novel was praised for its strong moral message. An 1847 review in *Godey's Lady's Book* praised the books "moral charms" in "assail[ing] the secret infidelity which is the bane of modern society."[46] Likewise, the same year a reviewer for the *American Agriculturalist* hailed its "moral bearing [as] excellent." Several decades later, an article in the *Christian Recorder* made reference to the novel's continued sway: "Who . . . can wonder that the French infidel, Compte de Charney, who spent months in the care and study of a delicate flowering plant, was led by its influence to believe in its Maker?"[47] The novel left an enduring impact on American letters. Mark Twain satirized the book's sentimentalism at the end of *The Adventures of Huckleberry Finn* (1884) when Tom Sawyer demands that the imprisoned Jim plant a flower and water it with his tears.[48] And adulatory reviews of the novel continued to appear into the 1890s.

Hawthorne himself most likely read *Picciola* in the 1840s and would certainly have been aware of it.[49] Moreover, *The Scarlet Letter* begins with Hester's release from prison and lingers over a rose perched at the prison door. For many of Hawthorne's readers and contemporaries, this rosebush would have resonated with Saintine's story about Charney's journey to redemption. But unlike Saintine's prison flower, Hawthorne's rose defies a neat moral lesson for the reader. The rose at the prison door is explicitly aberrant and hard to pin down: "Whether it had merely survived out of the stern old wilderness, so long after the fall of the gigantic pines and oaks that originally overshadowed it,—or whether, as there is fair authority for believing, it had sprung up under the footsteps of the sainted Ann Hutchinson, as she entered the prison-door,—we shall not take upon us to determine" (46). Here Hawthorne makes a point of disavowing any particular way of reading the rosebush. It has either "merely survived" changes in the land or is explicitly linked to Hutchinson. Sacvan Bercovitch describes how the narrator's refusal to choose leaves the meaning willfully indeterminate, producing a symbolic tension that absorbs conflict in a move toward pluralism.[50]

The rose's conspicuous presence on the threshold of the tale engages the reader familiar with botany as a common instructional trope but pays surprising dividends when the one character who directly compares herself to the flower is Pearl, whose self-identification powerfully resists conventional understandings of flowers as a means to female socialization. When Hester goes to Governor Bellingham's to retain custody over her daughter, the pastor John Wilson demands of Pearl, "'Can thou tell me, my child, who made thee?'" Her irreverent response shocks the room: that "she had not been made at all, but had been plucked by her mother off the bush of wild roses, that grew by the prison-door" (99).

Sacrilegious to be sure, this recourse to the botanical shifts the reader's attention from scripted catechism to the less predictable relationship between humans and plants. For Hawthorne, Pearl's identification with the wild rosebush by the prison door throws into relief the common analogies to plant growth through which authority is deployed. In this instance, the rhetoric of cultivation as socialization, so commonly relied upon as a metaphor for child-rearing, is volleyed back at the arbiters of

Pearl's fate. (Two years later, in *Uncle Tom's Cabin*, an equally pious Aunt Ophelia asks Topsy an identical question and receives a similarly shocking reply: "I spect I grow'd.")[51]

Hawthorne perverts the discourse of female socialization to challenge the role of plants as an ideology for manufacturing behavior. By layering the provenance of the rosebush, he renders acculturation a matter of accretion and exchange, rather than regulation and discipline. In doing so, he draws on the reality of plants as he and his nineteenth-century readers knew them: as scientific specimens, global commodities, and weeds, which were transplanted and hybridized with great regularity, yet whose development was not determined by human actions alone.

Pearl, of course, is no shrinking violet, and to any extent that she is socialized to a set of conventions, it happens off the page after she goes to Europe. Here, though, she externalizes and turns on its head the logic of cultivation within the Puritan context. The wild rosebush from which she declares herself plucked defies Calvinist views of sin and indeterminacy. Holly Blackford reads Pearl and Topsy as children who reflect the immorality of their communities at large.[52] Yet there is nothing reflective in Pearl's description of her origins. Instead, this botanical affinity shifts attention away from an exclusively human community in the discussion of identity—both individual and civic. In both scenes, the organic description of growth lies outside the bounds of normative ideology.

Governor Bellingham's garden further complicates horticultural cultivation as a process of stabilization. As Pearl and Hester wait for the governor in his house, the narrator notes that Bellingham had designed the home after English estates and gardens, though "the proprietor appeared already to have relinquished, as hopeless, the effort to perpetuate, on this side of the Atlantic, in a hard soil and amid the close struggle for subsistence, the native English taste for ornamental gardening" (94). What makes the description here noteworthy is its emphasis on the way in which the vegetation is distinctly not under full human control: "Cabbages grew in plain sight, and a pumpkin vine, rooted at some distance, had run across the intervening space, and deposited one of its gigantic products directly beneath the hall-window; as if to warn the Governor that this great lump of vegetable gold was a rich an ornament as New England earth would offer him" (94). The vegetation does not stay where it is planted, as the pumpkin vine "rooted at some distance" makes clear in its migration across the lawn. Even as Bellingham hopes to discipline Pearl and Hester, and as much as he would like to found a Puritan settlement on stable foundations, the very seat of his government suggests transplantation and mobility.

Vegetable growth, Hawthorne stresses, is hard to control, especially when weeds stand in for aberrant desires and sinful proclivities. In *The Scarlet Letter*, most famously, the "ugly weeds" that, according to Chillingsworth, grow out of a sinner's heart "typify, it may be, some hideous secret that was buried with [the dead man]" (114). Yet for all that the analogy serves the moralist, the role of unruly vegetation for Hawthorne was less clear-cut. In a journal passage that relates to his own experience gardening, he reflects,

> Why is it, I wonder, that Nature has provided such a host of enemies for every useful esculent, while the weeds are suffered to grow unmolested, and are provided with such tenacity of life. . . . What hidden virtue is there in these things, that it is granted them to sow themselves with the wind, and to grapple hold of the earth with this immitigable stubbornness, and to flourish in spite of obstacles, and never to suffer blight beneath any sun or shade, but always to mock their enemies with the same wicked luxuriance! It is truly a mystery. There is a sort of sacredness about them. Perhaps, if we could penetrate Nature's secrets, we should find that what we call weeds are more essential to the well-being of the world than the most precious fruits or grains. This may be doubted, however; for there is an unmistakable analogy between these wicked weeds and the bad habits and sinful propensities which have overrun the moral world; and we may as well imagine that there is good in one as in the other.[53]

It is a radically suggestive passage before the final reactionary turn, where Hawthorne bestows the familiar association between tares and sin. Only when the weeds become the subject of a precise analogical comparison to humans do they become imbued with negative associations. First, however, Hawthorne's description of weeds goes a long way toward cautioning us against reading the plants in his fiction in limited metaphoric ways, even though he makes this easy—too easy—for us to do. For a writer drawn to the ambiguous, the "unmistakable analogy" is a hollow pronouncement fixing these plants in service of a prescriptive moral register. Hawthorne's fiction exposes the limitations of this analogy by celebrating a natural world indifferent to human standards and desires. In the notebook passage above, weeds register as an alternative to the gardener's order and inspire because they are able "to sow themselves"—one hears Shelley's West Wind—and because they suggest that the cultivator does not own the soil.[54]

Such a notion challenges the idea of identity grounded in control of the landscape, and the supposed consensual politics of *The Scarlet Letter* thus appear in a different light when the novel's references to cultivation are considered. Hester can be taken as the embodiment of the consenting individual, and Pearl, who refuses to cross the stream in the forest unless her mother returns the scarlet A to her chest, a model of social enforcement. But if we place Pearl in the context in which she places herself—that is, in relation to a nature not under the gardener's thumb—we see her reversal of the parent-child disciplinary script. Most readings of the novel focus on Hester's relationship to the community at large rather than her relationship with her daughter.[55] "Impish" and inscrutable as she is, Pearl is harder to place, especially as she ends up knowing "that unknown region" across the Atlantic. Viewed from a generational perspective—"my children . . . shall strike their roots into unaccustomed earth"—Pearl's mobility challenges an exceptionalist view of America. By going abroad Pearl is freed from the community's edicts. Attending to his contemporaries' desire to legitimate consensus-building socialization in horticultural pursuits, Hawthorne identifies Pearl in relation to the uncontrollable and mobile aspects of the landscape.[56]

In this regard, Pearl neither conforms to the Puritan community nor is structured by her resistance to it. As Lee Edelman writes, "Politics, however radical the means by which specific constituencies attempt to produce a more desirable social order, remains, at its core, conservative insofar as it works to *affirm* a structure, to *authenticate* social order."[57] Pearl's location abroad at the end of the novel cannot be specifically plotted, and in this sense she occupies a space outside the American political scene, where even those who disagree still invest faith in the political system itself.[58] Although Chillingsworth bequeaths her property in America and England, she demonstrates no attachment to either except for her mother. If the scarlet letter marks Hester all the way to the grave, letters bearing "armorial seals . . . unknown to English heraldry" close Pearl's story (240). In this sense, the "flush and bloom of early womanhood" for Pearl is not, as it is for *Hobomok*'s Mary Conant, a question of civic identity tied to the landscape (239), or even a question of civic identity. Hester may be a model for external civil obedience, but she does not pass this on to her daughter. Pearl is more like the wicked but flourishing weeds whose growth cannot be trained into a prescriptive role within the American social order.

Something There Is That Doesn't Love a Wall: *The House of the Seven Gables*

In *The Scarlet Letter* Hawthorne illustrates how the mobility of plants posed a challenge to ideas of social cultivation. But plants' tendency to move also posed challenges to ideas of fixed borders and proved relevant to debates about property. Hawthorne brooded over this topic across his career, but nowhere did he engage with the meaning of property in a more sustained fashion than in *The House of the Seven Gables*, which revolves around a series of contested land claims. As many critics have pointed out, themes of ownership and identity preoccupy the novel, particularly through the Pyncheon's dubious claims to both the house in Salem and the Waldo territory in eastern Maine. The correlation between fixed settlement and land ownership was a deeply entrenched ideology that played an active role in Native American dispossession and resistance to English colonial rule. Mark Rifkin has lucidly illustrated how Hawthorne "mount[s] a critique of the state that does not so much seek to dismantle it as to install a regime (particularly of property) more consonant with an implicitly Lockean standard."[59] The novel's ending famously reifies property as the basis for social relations, as even the revolutionary Holgrave abandons his Fourier-inspired principles to live in the Pyncheon country estate once he and Phoebe are partnered. It is certainly a vision that upholds lineage: the expanded Pyncheon clan sallies forth to take up a comfortable residence upon another tract of land they inherit through family. Holgrave and Uncle Venner get absorbed into a new domestic order that is based upon, as Holly Jackson suggests, a reorientation of kinship and nationalism along the lines of race rather than class.[60]

Yet if the new concept of the family at the end of the novel makes race a kind of natural property, the depictions of plants throughout the novel trouble the coherence of such formulations. Nature itself, represented in the botanicals that surround the house, is shown to be mobile and hybrid. As we've already seen, Hawthorne wrote at a moment when the circulation of plants and implementation of new forms of garden technology were fundamentally changing the way that people conceived of their control over nature. His novel destabilizes nature as an essentialist category to show how these changes were shifting the correlations between property and identity.

As we saw in the first chapter, one foundation of American literary nationalism in the late eighteenth century and early nineteenth was the claim of natural distinctiveness within the country's borders even as these borders continued to expand. By 1850 that notion of geographic essentialism held less currency, as the heavy circulation of plants and rise of commercial nurseries made the incorporation of foreign plants a common household practice, bringing plants from around the world into the most intimate of spaces. What Harriet Ritvo notes of midcentury England was similarly true of many areas of the United States: as a result of "the nineteenth-century democratization (or at least bourgeoisification) of horticulture, it became possible for any middle-class hobbyist to construct a miniature empire in the back garden."[61] The popularization of gardening as a healthful part of domestic life—espoused by arbiters of the home like Lydia Maria Child and the Beecher sisters—contributed to the growing influx of nonnative plants. Hawthorne's uncle regularly received shipments of trees from around the world, and Hawthorne himself would order trees from England to plant at his homestead later in his life.

The key point here is that while the unique qualities of the American landscape remained a central component of many literary representations of the nation, by the 1840s horticultural manuals, nursery catalogues, and articles in popular magazines invited the influx of plants from different parts of the globe, particularly as improvements in greenhouse design and heating technology made the cultivation of plants from other climates possible. By the postbellum period, it was not uncommon for middle-class families in the Northeast like the Dickinsons to have a greenhouse attached to their house. In the domestic space, transplantation was encouraged as an enterprise good for the health of the cultivator and for the advancement of horticulture. The increasing technological involvement, scientific experimentation, and popular practice of plant cultivation challenged simple notions of botanical nationalism and primed readers to see nature in a new light not as essence but as experiment. Hawthorne renders plain anxieties about such experiment in "Rappaccini's Daughter," and in *The House of the Seven Gables* he returns to the ways in which such activity could threaten patriarchal stability.

Early horticultural institutions often projected a fantasy of control. As horticultural journals and magazines made clear, the goal of both professional and amateur gardeners was plant improvement. Improvement, in turn, was increasingly defined by horticultural societies, emerging arbiters of cultivation, as a matter of exacting precise control over the soil. As the president of the Massachusetts Horticultural Society argued in 1848, the goal of every gardener is to appreciate "that he is not merely the tenant, but in a proper sense, the lord of the soil." Recalling a time "when without the light of science, the old worn out systems and routines of cultivation, were handed down from sire to son, and from generation to generation," the president extolled "our happy lot to live at a period when a new era has commenced—when the most distinguished and learned men of our age are joining hands to advance the cause of the cultivator—when chemistry, geology and the mechanic arts have come up to his aid."[62] The narrative here—of darkness to light, of transitioning from timelessness to a new era— makes progress a matter of scientific disciplines and asserts the male cultivator as godlike figure. The argument from design gives way to the "cause" of the farmer or gardener in a logic that sets the idea of botanical improvement against the idea of American nature as timeless and essential. To manage the landscape in the name of improvement meant relying on transatlantic networks of horticulture and botany. In this sense, exceptionalist narratives of national improvement coexisted with competing transnational realities.

Especially during his own transnational travels, Hawthorne scorned the rhetoric of mastery over the soil. In his capacity as U.S. consul in Liverpool in the mid-1850s, he rejected what appeared to him to be a manipulative, totalitarian planting regime. He chronicles in his "English

Notebooks" the estate gardens he visits in great detail and is struck by prominent horticultural technology. For example, he describes Poulton Hall as a landscape that has been artificially coaxed to "get everything from Nature which she can possibly be persuaded to give them, here in England."

> Peaches and pears growing against the high brick southern walls,—the trunk and branches of the trees being spread out perfectly flat against the wall, very much like the skin of a dead animal nailed up to dry, and not a single branch protruding. . . . The brick wall, very probably, was heated within by means of pipes, in order to reinforce the insufficient heat of the sun. It seems as if there must be something unreal and unsatisfactory in fruit that owes its existence to such artificial methods. Squashes were growing under glass, poor things![63]

Hawthorne's sympathetic response to the poor fruits and vegetables of Poulton Hall reflects a palpable anxiety about the means used to grow them. "Artificial methods" here deaden the plants, so that the trees look denatured: a kind of living death. On another occasion while touring the countryside, Hawthorne describes "a row of unhappy trees . . . spread out perfectly flat against a brick wall, looking as if impaled alive, or crucified, with a cruel and unattainable purpose of compelling them to produce rich fruit by torture" (257). Hawthorne's representation of cultivation here renders it violent, and while he is more or less charmed with the English countryside, he finds that nature here cannot be distinguished from engineering efforts. "The wildest things in England are more than half tame" (114), he writes in his notebook, repeating himself more strongly later that "the landscape was tame to the last degree" (185). The deadening sensibility that Hawthorne describes here is a theme that animates much of his earlier fiction.

Hawthorne's stories expose the extraordinary effort required to force plants to respond to human desires. As any experienced gardener knows, plants rarely behave according to a set plan, nor can climate, insects, or animals be fully disciplined. Gardeners and horticulturalists could decide where to introduce nonnative plants, but not necessarily where they wound up. Plants, along with fungi and insects, often moved in unpredictable ways as they pursued their own interests.[64] In his fiction, Hawthorne recuperates the mobility and autonomous actions of plants to challenge essentialized notions of identity and property.

Hawthorne's depiction of gardening and gardeners in *The House of the Seven Gables* at times reifies the classed and gendered discourse that surrounded flowers, fruits, and vegetables in midcentury novels and periodicals. As Philip Pauly explains, the cultivation of fruit trees in the first half of the nineteenth century was considered a gentlemanly pursuit because "vegetables were plebian, flowers were effete, ornamental shrubs were largely unavailable in North America, and timber trees were either too common or too slow-growing."[65] Judge Pyncheon is, among other things, a horticulturalist who "produc[es] two much-esteemed varieties of the pear"; Phoebe Pyncheon is a young country cousin whose femininity is associated with her blooming cheeks and skill with flowers; and in contrast to Phoebe's cultivation of "aristocratic flowers," Holgrave tends to the garden's "plebian vegetables."[66] Yet for all this careful cultivation that reinforces lines of gender and class, the land itself and the plants that grow there reject such assignations.

At first it may appear that they do the opposite. The sardonic narrator describes with pleasure the many ways that the land registers the immorality of the ancestral Pyncheons' land grabbing: the water from Maule's well famously grows brackish and "productive of intestinal mischief to those who quench their thirst there" (10) and the rosebush that Phoebe spies from her window on her first night in the house is, upon closer inspection, blighted at its core. Such easy symbolism seems to be a simple moral judgment until we consider that Hawthorne is also engaging with contemporary theories about soil fertility and crop rotation.

From the late 1830s horticulturalists and agriculturalists fiercely debated the theory of the execratory powers of plants. One strongly held argument for crop rotation rested on the idea that plants excreted substances into the soil that were poisonous for future generations of the same plant species. An 1834 article from the *Genesee Farmer and Gardener's Journal* describes it thus:

> It has been very generally noticed, that the soil in which any particular plant has grown, and into which it has consequently discharged the excretions of its roots, is rendered noxious to the growth of plants of the same or of allied species. . . . The whole theory depends upon the fact,

that all plants succeed badly upon lands which have lately borne crops of the same species as themselves, or even of the same genus, or of the same family. This effect . . . arises from a corruption of the soil, by the intermixture of vegetable excretions given out at the root, which excretions are always more deleterious to plants of the same kind than to others. It is even ascertained that the excretions of some plants are beneficial to the growth of others of a different family.[67]

This theory was hotly debated among botanists and agriculturalists and was the impetus for many experiments into the relationship between roots and soil.[68] And the questions that horticulturalists were asking resonate with the questions that Hawthorne posits in *The House of the Seven Gables*: To what extent was soil a product of contemporary human activity? What traces of the past endure in the ground? And to what degree does a plot of cultivated land require change in order to stay vital?

Hawthorne draws on debates over soil quality to demonstrate that the ground was not an essentialized or uncontested reference point for identity. At a key moment in the novel Holgrave memorably exclaims to Phoebe that "under that roof . . . there has been perpetual remorse of conscience, a constantly defeated hope, strife amongst kindred . . . dark suspicion, unspeakable disgrace—all or most of which calamity I have the means of tracing to the old Puritan's inordinate desire to plant and endow a family. To plant a family! This idea is at the bottom of most of the wrong and mischief which men do" (185). Holgrave's comparison and his incredulous tone encourage the reader to dismiss the metaphor, part of his larger point about the obdurate and unjust nature of the Pyncheons' claim to the property. Yet the novel is stubbornly committed to botanical imagery and knowledge, and in the context of the way plants and property are depicted throughout, Holgrave's outburst bears closer scrutiny.

The Pyncheons' claim to the Salem property is rendered, by the narrator, as dubious as the deed claiming ownership of most of Waldo County. In the latter instance, however, the narrator elides the issue of Indian land rights and reifies the concept of land ownership as belonging to white settlers by virtue of their labor: "These [settlers], if they ever heard of the Pyncheon title, would have laughed at the idea of any man's asserting a right—on the strength of moldy parchments, signed with the faded autographs of governors and legislators long dead and for-gotten—to the lands which they or their fathers had wrestled from the wild hand of nature by their own sturdy toil" (18–19). Ownership here is a matter of people imposing their will over a discrete parcel of "wild nature" that is, by the very logic of its being subdued, rendered theirs.

This passage functions to excise Native claims to the land twice: once in the legal form of the deed, "confirmed by a subsequent grant of the General Court" (18), and once in the form of the settlers who lay claim through the Lockean justification of toil. The elision of Native claims in *The House of the Seven Gables* functions along the lines of many of Hawthorne's other stories to, as Derek Pacheco writes, "reproduc[e] the cultural myth of the vanishing Indian, a fantasy of a race destroyed not by the policies of Anglo-America, but by the movement of time itself."[69] In *The House of the Seven Gables*, Hawthorne explores the logic of land grabbing, where to settle—by force—is to own. Yet he goes beyond that to explain how force is superseded by cultivation, whose cyclical nature helps complete amnesiac erasure of earlier land use.[70] Hawthorne sug-gests how acts of land appropriation were legitimated through the linked rhetoric of horticulture and domesticity.

It is this ideology of horticultural progress to which Hawthorne turns an eye at once romantic and skeptical in *The House of the Seven Gables*. The private garden at the back of the house is a space that seems far removed from the economic relations modeled in Hepzibah's cent shop, or indeed in the crowded public spaces like the parade that passes by the Pyncheon house. Yet part of the novel's work is to put the garden's romantic associations in tension with the kind of rational, classifica-tory rhetoric that filled horticultural registers and botany textbooks and shaped public discourse about place. On the one hand, the novel is full of botanical language that seeks to explain moral truths on an abstract plane. For instance, "The act of the passing generation is the germ which may and must produce good or evil fruit in a far-distant time; that to-gether with the seed of the merely temporary crop, which mortals deem expediency, they inevitably sow the acorns of a more enduring growth, which may darkly overshadow their posterity" (6). On the other hand, these seeds serve a literal function in the text as well. Holgrave plants the seeds he finds growing in a garret over one of the seven gables, "trea-sured up in an old chest of drawers by some horticultural Pyncheon of days gone by." The result of his experiment, "testing whether there were still a living germ in such ancient seeds," produces "a splendid row of bean vines, clambering, early, to the full height of the poles, and arraying them, from top to bottom, in a spiral profusion of red blossoms" (148). By introducing empiricism into the text through Holgrave's horticultural experiment, Hawthorne uses the practice of cultivation to legitimate the story's more figurative views about sin and redemption. The scarlet blossoms on the bean vines attract hummingbirds, which Hepzibah understands in relation to Clifford's freedom: "And it was a wonderful coincidence, the good lady thought, that the artist should have planted these scarlet-flowering beans—which the hummingbirds sought far and wide, and which had not grown in the Pyncheon garden before for forty years—on the very summer of Clifford's return" (148). Nature here endorses Clifford's return, but only as a result of Holgrave's deliberate cultivation experiment. Through Hepzibah's thought process Hawthorne here renders transparent how horticultural enterprises shaped the very definition of nature upon which romantic notions of justice rested.[71]

Alice's posies likewise serve to highlight the relationship between deliberate, historically located acts of cultivation and the romantic authenticity of timeless Nature. Through recurrent description of these flowers, Hawthorne exposes how the realities of domestic gardening were serving as new foundations for considering nature not as some fixed essence but as the product of historically contingent human activities and the agencies of plants themselves. Like the rosebush in *The Scarlet Letter* or the hothouse flowers that Zenobia wears in her hair in *The Blithedale Romance*, these flowers growing out of the house call immediate attention to their role as symbols. In the first chapter the narrator enshrines them in a frame narrative: "Tradition was that a certain Alice Pyncheon had flung up the seeds, in sport, and that the dust of the street and the decay of the roof gradually formed a kind of soil for them, out of which they grew, when Alice had been long in her grave. However the flowers might have come there, it was both sad and sweet to observe how Nature adopted to herself this desolate, decaying, gusty, rusty old house of the Pyncheon family" (28). Situated on top of the house, the flowers are botanical spectacles that straddle not only the angle between the two front gables but the line between human volition and plant agency. The sentence structure both displays and then effaces the fact of their anomalous growth: first the narrator points out that "Alice Pyncheon had flung up the seeds," but then quickly dismisses this fact by adding the next sentence that begins with, "However the flowers might have come there" to pursue a point about Nature's sympathetic relationship to the house. Unlike a flower growing under glass or underfoot, the flowers here grow outside human management, yet the domestic space provides the very substrate in which they take root. Perched atop the house, they are, unlike the plants in the garden, visible to the street and to any passerby and thereby rendered as legible signs for what lies hidden within. Yet what they most reveal is their own ability to thrive in unexpected if not unaccustomed earth.

Tensions between human and organic agency recur at the end of the novel in the chapter entitled "Alice's Posies." These flowers, of course, constitute a powerful symbol for the impending union of Phoebe and Holgrave, which in turn absolves the present Pyncheons of the sins of their forebears. The fact of their coming into bloom after Judge Pyncheon dies "seemed, as it were, a mystic expression that something within the house was consummated" (286). The late reference to these flowers echoes their description at the start of the tale and places them within a metacritical commentary about romantic Nature as a grounds for authenticating social relations. Consider their return near the end of the novel: "One object, above all others, would take root in the imaginative observer's memory. It was the great tuft of flowers—weeds you would have called them a week ago—the tuft of crimson-spotted flowers, in the angle between the two front gables. The old people used to give them the name of Alice's Posies, in remembrance of fair Alice Pyncheon, who was believed to have brought their seeds from Italy. They were flaunting in rich beauty and full bloom, to-day" (286). The retelling here is curious because we have learned about these flowers already, very early on, and have already been told that they were not weeds, "but flower shrubs, which were growing aloft in the air" (28). Here the nomenclature—weed versus flower—is shown to be subjective. Moreover, as in their first introduction, when "tradition" guides their meaning, the posies are contextualized by the names the "old people" give to them (28, 286). Hawthorne is emphasizing how much the flowers have become part of a collective mythos: rather than give the reader the image as a symbol to interpret, he gives it as a symbol that has already been interpreted. And we are pointedly made aware of that fact. While this mediation is a narrative strategy that recurs throughout the novel, it is especially significant in relation to the botanical descriptions because they ground so many of the novel's moral claims, such as the eventual absolution of ancestral sin. Alice's posies literalize how those moral symbols are embedded in the specific context of domestic gardening. Gardening, in turn, is shown to be an inherently political activity that both reshapes the land and undergirds the horticultural rhetoric justifying such activity.

Horticulture was significant in this regard for how it affected power dynamics in both domestic and public realms. Hawthorne was writing at a moment when conversations about planting were extending beyond the home garden to the level of communal space. Alice's posies are poised between the domestic and public spaces and gesture to the way that foreign plants helped to substantiate domestic landscapes. Sprouting out of "seeds from Italy" on the roof of a Salem residence, the posies offer a particularly powerful example of how "domestic metaphors of national identity are intimately intertwined with renderings of the foreign and the alien, and that the notions of the domestic and foreign mutually constitute one another in an imperial context."[72]

Transplantation, as we saw in chapter 1, could serve as a metaphor for naturalizing settler colonialism. But biotic movement could not always be carefully controlled, and Hawthorne recognized the power of unanticipated consequences to flout patriarchal authority. Part of what is striking about Alice's posies is that they grow in an out-of-reach place, spread (so it is rumored) by human intention but otherwise outside the realm of human touch. Most studies of *The House of the Seven Gables* approach it from the perspective of its commentary on social control. Maule is a mesmerist, as is Holgrave to an extent (a quality shared, as Samuel Coale points out, with Westervelt, Alymer, Rappaccini, and Hollingworth), and the prose style can be said to enact the same kind of control over the reader.[73] Yet for all that this is apparent in the novel, so too are the moments when Hawthorne concedes agency to a nature beyond the realm of the mesmerist/medium dichotomy. For the empiricist reader, the Pyncheon yard is full of evidence of nature's own dynamic, undisciplined processes.

Environmental fluidity, in other words, could unsettle organic claims to control. In addition to the cultivated flowers and vegetables that appear in the garden, the house is surrounded by vegetation such as "an enormous fertility of burdocks, with leaves, it is hardly an exaggeration to say, two or three feet long" (27). The mention of burdock is significant; the *Home Journal* in 1856 characterized the plant as a common weed with flowers that "changed into burs" and then "reached out their thorny fingers and grasped the passers-by . . . and the seeds flew out on the wind to seek lodging places, where another year a new crop should find foothold and sustenance."[74] Hawthorne's naming of the burdock suggests familiarity with a plant notorious for its ability to spread itself by adhering to humans or animals that come in contact with it and using their mobility for its own.

The squash blossoms in the back garden are likewise part of a process that connects them to an organic logic beyond the garden purview. In a passage shortly before the description of the hummingbirds that delight Clifford, Hawthorne describes the arrival of a swarm of bees: "Thither the bees came . . . and plunged into the squash blossoms, as if there were no other squash vines within a long day's flight, or as if the soil of Hepzibah's garden gave its productions just the very quality which these laborious little wizards wanted, in order to impart the Hymettus odor to their whole hive of New England honey" (147). That bees come from all over New England to the garden highlights mobility not under the control of the cultivators, although the narrator seeks to claim "why the bees came to that one green nook in the dusty town" explaining that "God sent them thither to gladden our poor Clifford" (148). This impulse toward legibility in the context of the narrative nonetheless betrays a strong recognition that nature, while responsive to human activities, operates according to principles outside human control. The *Putnam's* author of "A Chat about Plants" echoes this claim: "The poet of old already has taught us, that you may drive out nature even with the pitchfork, and yet she will ever return. A few years' neglect, and how quickly she resumes her sway! Artificial lakes become gloomy marshes, bowers are filled with countless briars, and stately avenues overgrown with reckless profusion. The plants of the soil declare war against the intruders from abroad, and claim once more their birthright to the land of their fathers."[75] The sensibility here is akin to Hawthorne's suggestion in *The Scarlet Letter* and *The House of the Seven Gables* that plants defy the property model upon which antebellum social relations functioned. Before Phoebe returns to the Pyncheon property, the "neglected yard" is described as "now wilder than ever, with its growth of hogweed and burdock" (293). Upon her arrival she discovers that "the growth of the garden seemed to have got quite out of bounds: the weeds had taken advantage of [her] absence, and the long-continued rain, to run rampant over the flowers and kitchen vegetables" (299). Without constant intervention, the garden plants grow according to their own logic. What is natural here is not the carefully cultivated plot but the weeds that grow in absence of human labor to carefully weed, prune, and manage the area.

It is in these kinds of moments that Hawthorne's plants convey a more ambivalent understanding of property. "Something there is that doesn't love a wall," Robert Frost would write in 1914 in a poem about how property structures relations between neighbors. Hawthorne likewise tapped into the ways that plants ignore property boundaries to register deep ambivalence about cultivation.

* * *

The nineteenth century is often characterized as a period that gave rise to divisions, partitions, and borders that still inform academic practices and knowledge production today. The professionalization of science, the increased division of labor associated with industrial capitalism, and the attendant transformations and divisions of landscape: these are legacies of an age of empire that continue to bear on our contemporary political and environmental landscape.[76] But Hawthorne's stories remind us that these changes in the land did not simply spell patriarchal domination. Writing at time when the Concord grape was discovered by accident in a corner of a neighbor viticulturalist's yard, Hawthorne registers the subversive possibilities of plants to challenge scientific and social efforts to control them as commodities and specimens, unsettling some of the period's most trenchant binaries in the process. When Holgrave mocks the Pyncheon desire to "plant a family," we see Hawthorne's evocation of plants' ability to pursue reproductive strategies outside the realm of human control. Such an ability, he understands, challenges the dichotomy of possessor and possessed, of agent and object, that is inseparable from the concept of ownership.

Acknowledging that plants pursue their own strategies invites a number of further questions about their ontology. As the next chapter shows, plants' abilities to travel on their own, to reproduce and survive in unlikely conditions, and to react to stimuli were just some of the behaviors that sparked curiosity in the minds of nineteenth-century scientists and home gardeners. From the late eighteenth century into the middle of the nineteenth, scientific theories of plants' sentience attracted the attention of journalists and gardeners. Did a flower that folded up at night feel? Did a plant root that crossed the yard to find water think? Scientists struggled to characterize plant behavior and to provide language that explained what they were seeing without giving plants too much credit. They vacillated on the edge of certain ideas and struggled with the implications of their own language. In the process they influenced the way a generation of gardeners thought about the plants growing in their gardens. One of the best ways to see the tensions inherent in this language is to examine poetry alongside scientific and popular writing. As Emily Dickinson's experimental verse shows, poetry proved an especially powerful medium for exploring the limits of language when presented with the liveliness of plants.

3

Botanical Agency

Science, though a very great and learned lady, does not yet
know everything. Her elder sister, Poetry, often sees further
and deeper into things than she does.
—*All the Year Round*, "Have Plants Intelligence?," 1870.

In 1870 the American *Independent* ran an article from Charles Dickens's
magazine *All the Year Round* titled "Have Plants Intelligence?" The provocative question in the title was designed to spark intuitive negative
responses, but the paragraphs that follow rehearse a clear argument in
the affirmative. Life itself "presupposes in its possessor, whether animal
or vegetable, a faculty of sensation that administers to its happiness, and
that may consequently administer to its suffering," the author argues.
This meant that plants experience pleasure and pain.

Although the author suggests that the scientific community had dismissed the notion of feeling plants, in fact naturalists had been asking
and debating this issue for decades, studying plants like the carnivo-
rous flytrap that caught prey and the "sensitive" mimosa that shrank
upon touch. By the mid-nineteenth century a number of scientists
believed that plants could at least feel, if not think, and their findings
were received by audiences whose own experiences cultivating plants
had allowed them to observe a stunning array of plant behaviors. In the
garden, the parlor, and the greenhouse, plants' living qualities became
an object of fascination and raised questions like the one the article
poses rhetorically. How to make sense of the behaviors of plants? Did
they have an inner life? It could certainly appear so, though so much
remained a mystery.

If the scientific community was not as categorically opposed to the
idea of plant feeling as the author implies, the article's turn to poetry
suggests literature might excel at exploring the inherent difficulties in
representing this other form of life. Instead of cold science, literature's

different approach to feeling, sensing, and knowing, and above all its careful investment in language and imagination, allowed for a distinct approach to understanding other forms of life, even life that strained or surpassed comprehension in language. In short, literature could access knowledge about plant life that scientific empiricism could not. Such romantic reactions were taking place across the natural sciences at the time. As Sari Altschuler has shown, nineteenth-century members of the medical community turned to literature to better understand the mysteries of bodies, diseases, and environments.[1] Could literature likewise yield insight or appreciation for the liveliness of plants? This chapter explores this question by way of popular writing on plants and the poetry of Emily Dickinson, whose passion for gardening and sophisticated grasp of contemporary natural history had a profound effect on her experimental verse.

Dickinson was, by all accounts, a skillful and dedicated gardener. Throughout her isolation at her parents' house on Main Street in Amherst, Dickinson continued to raise plants, arrange bouquets, and send cuttings to distant friends. As a student she scouted for new flowers to press into her bound herbarium, and in winter, to keep plants warm, she brought them into the conservatory built against the southeastern wall of the house. The structure allowed her to cultivate tropical plants that could not otherwise survive the New England climate. Dickinson knew firsthand how fickle plants could be, and how fragile, and she had skill in growing plants that were not native to the area or suited to the climate. And her many botanical poems take on new meaning when read alongside the circulation of plants across the nineteenth century that made foreign flora accessible to American home gardens.

Sentiment and science overlapped in rendering plants intimately strange to nineteenth-century gardeners.[2] As conventional sentimental tokens, flowers conveyed emotional significance. Their cultivation served, as we saw in the last chapter, as a popular metaphor for education and socialization. As commodities and specimens, meanwhile, plants were valued within the economic and scientific systems in which they circulated. This overlap of meanings often brought into focus the lively and unpredictable materiality of plants, which could challenge human efforts to understand or control them. As nineteenth-century naturalists moved plants around the globe and attempted to cultivate them in new climates, plants sometimes failed to thrive in new environments or took root in unexpected ways, and their existence challenged the values that humans assigned to them.

Across her poems and her letters Dickinson dwelled on how the creative energy of the botanical realm might escape, challenge, and in some ways reorganize human-centric designs. In this sense, she departed from dominant theories of natural philosophy that elevated human consciousness above other forms of life, aligning herself instead with an emergent scientific discourse about plant feeling. Whereas the many theories of life in the nineteenth century—like the great chain of being or argument from design—tended to see the world as an orderly and stable hierarchy with humans at the top, Dickinson was among those who found in the plant realm another possibility: life whose very nature is collaborative, decentralized, and communicative with other environmental agents in ways that human actors cannot anticipate or control.

Such theories about the organization of plant matter inevitably have political and cultural consequences.[3] If plants are vital, sensible, and mobile, they cease to simply reflect the human values projected upon them. Their autonomy is both difficult to imagine and politically charged, for it elicits an awareness of the limits to human control—and not only control of plants themselves. Dickinson was among a growing number deeply interested in plant material for how its creative forces might instruct human life, not only for the moral cultivation of the individual, but also for the organization of society. The riot of life Dickinson depicts in the garden is therefore not only about allegory, subjectivity, or romantic aesthetics. Rather, because botanical vitality was recognized as being distinctive from human biology while at the same time unnervingly familiar, plants could challenge assumptions about the hierarchies that governed human affairs.

This chapter focuses on two often overlooked dimensions of cultivating plants at home—their circulation and their liveliness. From her garden, Dickinson had occasion to see how the global traffic in plants transformed local environments and to study the ways plants moved on their own. The first part of the chapter takes up this issue of plant mobility, considering the ways that Dickinson's poetry reflects the circulation of plant matter during her lifetime. The second part shows how concepts of plant irritability, sentience, and intelligence were taken up in nineteenth-century botanical culture, examining how Dickinson engaged the possibility of an active, feeling natural world. These ideas posed a linguistic as well as an ontological challenge to popular understandings of life, and taken together, these sections address how poetry's formal possibilities provided a medium for exploring the limits of languages to characterize other forms of life.

Plant Mobility

Home gardening in mid-nineteenth-century Amherst was far from a domestic enterprise.[4] With the influx of plants from abroad, nineteenth-century botanists and home gardeners were increasingly aware that the data they could empirically collect about their plants might be influenced by forces beyond their local purview. Dickinson's own poems encourage us to consider plant and human behavior on a global scale. Rejecting the notion that the natural world could serve as a stable ground for reifying social order, Dickinson's flowers, birds, and even her poles move. Tulips that grow at home are transplants from Asia, flowers wander, and trees do not stand still, as "When oldest Cedars swerve— / And Oaks untwist their fists—" (F882 A). In some poems, vast intervals compress into tight stanzas. The distance between western Massachusetts and the Kashmir region of the Indian subcontinent—well over six thousand miles—is one that few in Dickinson's lifetime would travel. The poet, however, traversed these miles within the poetic line. "If I Could Bribe Them by a Rose" (F176 A) begins,

> If I could bribe them by a Rose
> I'd bring them every flower that grows
> From Amherst to Cashmere!
> I would not stop for night, or storm—
> Or frost, or death, or anyone—
> My business were so dear![5]

Dickinson's poetry often spans such distances, bringing places as diverse as Amherst and Cashmere, or New England and Santo Domingo, into the same imaginative sphere. The generative energy of this dimension of her writing has led to a turn in criticism that has unmoored Dickinson from the fixed radius around Amherst and even the United States. Dickinson's ability to "telescope" place, as Christine Gerhardt puts it, is often particularly notable in the floral language she evokes over the course of her poetic career, because Dickinson was keenly aware that flowers simultaneously filled the local garden and circumnavigated the globe.[6] This motion resists the explanatory power of conceptual categories like the local or national, and Dickinson's political engagement makes fuller sense when we consider how plants functioned within an international framework. Dickinson's poems can suggest how middle-class horticultural enterprises were facilitated by colonial botanical pursuits, and how the projects of the home gardener were tied to imperial designs.

One offshoot of colonial bioprospecting was the development of a commercial plant industry that catered to wealthy and middle-class citizens, providing a window onto the world's flora that could be browsed in the comfort of the home. Seed and plant catalogs developed rapidly across the nineteenth century, depicting new varieties from around the globe in increasingly lush and lavish visual depictions. Nurseries on both sides of the Atlantic advertised novel plant species for the home garden and frequently identified in their catalogs the source of the new species. At the Dickinson homestead, as Judith Farr has determined, the nursery catalogs included those by L. W. Goodell and B. K. Bliss.[7] Bliss's 1870 spring catalogue assigns each variety a "Native Country," and the list includes "France," "Mexico," "California," "East Indies," "Russia," "Chili [sic]," and "N.S. Wales," among many others.[8] More widely, such catalogs often identified origins in terms of not only nations but also continents, riverbanks, or mountain ranges. The 1862 London *Barr and Sugden Guide to the Flower Garden, &c.*, for instance, names as origins not only "N. America," "Canada," "SW Australia," "Africa," "West Indies," "E. Indies," "France," "China," and "Germany," but also "Himalaya," "The Levent [sic]," "Arabia," "Straights [sic] of Magellan," "Swan River," and the "Caucauses."[9]

In addition to seed and plant catalogs, news of regular international plant circulation was available to readers of horticultural periodicals. The proceedings of the Massachusetts Horticultural Society were regularly published in the *New England Farmer* and elsewhere, containing information on the society's new acquisitions from abroad, such as "samples of the grape vines, cherries and other fruits of the Crimea,—seeds of such forest trees as were considered valuable for economical purposes."[10] The regular requests for new seeds—particularly those perceived as potentially economically valuable—fill the annals of horticultural institutions. Indeed, the craze for international seeds frustrated those nurserymen like Joseph Breck who had to keep pace with changing tastes. In *The Flower-Garden; or, Breck's Book of Flowers* (1851), which Dickinson and her sister likely read, Breck speaks to the capriciousness associated with the availability of new imports: "We remember when Cape plants were the rage; . . . But in a few years these were thrown aside, and New Holland beauties supplanted them; to be succeeded by the flaunting, or shy and delicate, natives of South America."[11] Nursery culture by midcentury was tied into global networks of bioprospecting, and customers came to expect the annual arrival of new varieties from abroad.

Dickinson engages with the concept of bioprospecting as both a personal and general phenomenon. In "I Robbed the Woods" (F57 A) the speaker expresses mild approbation at her own plant collection in the woods: "I grasped—I bore away—" has a confessional tone, but the use of the first-person pronoun suggests a limited scale. The poem may even have been a critique of her brother's habit of taking specimen trees out of the nearby woods to plant on his own property.[12] In a second variation of the poem, the "I" is replaced by "Who." This shift transforms the implications for the plant collection, making it far more reaching. "Who Robbed the Woods—" (F57 B) might allegorize a larger exploitative operation:

> Who robbed the Woods—
> The trusting Woods?
> The unsuspecting Trees
> Brought out their Burs and Mosses—
> His fantasy to please—
> He scanned their trinkets—curious—
> He grasped—he bore away—
> What will the solemn Hemlock—
> What will the Fir tree—say?

Here the natural world is translated into a commodity: burs and mosses become "trinkets" to be carried off by an anonymous exploitative agent. The relationship between this version of the poem and the version engaging the personal pronoun suggests Dickinson's willingness to consider how the local, personalized collections of botanical specimens might fit within a larger historical paradigm.

"The Robin's My Criterion for Tune" (F256 A) also acknowledges environmental change in the local landscape. The poem is most often read as an exploration of a distinctive New England context, but it can also be read "slant" as ironizing the concept of biotic regionalism in an age of plant nurseries and imperial gardens.

> The Robin's my Criterion
> for Tune—
> Because I grow—where
> Robins do—
> But, were I Cuckoo born—
> I'd swear by him—
> The ode familiar—rules the
> Noon—
> The Buttercup's, my whim
> for Bloom—
> Because, we're Orchard sprung—
> But, were I Britain born,
> I'd Daisies scorn—(spurn—)
>
> None but the Nut—October
> fit—
> Because—through dropping it,
> The Seasons flit—I'm taught—
> Without the Snow's Tableau
> Winter, were lie—to me—
> Because I see—New Englandly—
> The Queen, discerns like me—
> Provincially—

The speaker of the poem associates with the robin, buttercup, and daisy because they share a habitat. As Christine Gerhardt argues, "The poem suggests that the speaker is an equal member of the region's biotic community, and that New England is defined by inter-locking cultural and natural systems." Seeing "New Englandly" is causally related to "growing" there, a clear articulation of place and of regional difference. And as Gerhardt notes, contemporaneous dis-courses of plant geography and birding helped to establish a sense of New England as a distinct place.[13]

But when the poem extends this sense of provinciality to the queen of England in its final lines, we might ask if seeing provincially means something more in this context. On the one hand the poem strongly links perception to place, suggesting that the speaker and the queen share a regional mode of perception based on the distinctiveness of the local biota. At the same time, however, the queen's metropolitan seat of power begs us to reconsider discerning provincially. In the con-text of the British Empire, it becomes a matter of understanding how the movement of objects and knowledge might link disparate regions.

Read this way, we might see this poem as wryly acknowledging the ways in which local ecological change was the product of envi-ronmental transformations on a global scale. By connecting the New England "I" to the British queen, the speaker establishes parity be-tween her own environmental regard and that of the symbolic head of an empire that had already claimed provinces across six continents. The biota of England in the mid-nineteenth century had been signifi-cantly augmented by colonial projects, and as Alan Bewell and others have noted, by the early nineteenth century, colonial natures—in the form of plants, minerals, birds—were "flooding" into Europe from around the globe on an unprecedented scale.[14] Natural history and economic botany were linked practices that facilitated this exchange of natural materials. New England was the recipient of colonial plants, technology, and cultivation practices, as the popularization of garden-ing among the middle class created a market for novel plants from abroad.

The "transport" that flowers could inspire thus might be perceived to be as much geographical as sublime. In "Some Rainbow—Coming from the Fair!" (F162 B), the arrival of spring in the local arena is her-alded as an international affair. The changing seasons invoke a sense of geographical compression from the start of the poem to its end:

Some Rainbow—coming from the Fair!
Some Vision of the World Cashmere—
I confidently see!
Or else a Peacock's purple Train
Feather by feather—on the plain
Fritters itself away!

The dreamy Butterflies bestir!
Lethargic pools resume the whir
Of last year's sundered tune!
From some old Fortress on the sun
Baronial Bees—march—one by one—
In murmuring platoon!

The Robins stand as thick today
As flakes of snow stood yesterday—
On fence—and Roof—and Twig!
The Orchis binds her feather on
For her old lover—Dons the Sun!
Revisiting the Bog!

Without Commander! Countless! Still!
The Regiments of Wood and Hill
In bright detachment stand!
Behold! Whose multitudes are these?
The children of whose turbaned seas—
Or what Circassian Land?

The poem forges a link between local natural phenomenon and military engagement elsewhere. Farr argues that the "near military formation" of the flowers "describes their aesthetic potency," and that the poem's international allusions are ultimately there to stress the "power of beauty."[15] In this reading, the focus is on aesthetic interpretation: the arrival of spring as a timeless, powerful force. But the poem's geographical references are, in a way, constitutive of its aesthetic intensity.[16]

The poem begins with clarity of vision about the landscape that slowly gives way to uncertainty. The declaration that "I confidently see!" in the first stanza concedes to questions of origins in the last. Moving from Cashmere to Circassia over the course of four stanzas to characterize spring, the poem conflates beauty not with truth, as a Keatsian romantic might, but with conflict. The allusion to Circassian Land invokes contested territory during the ongoing Russian-Circassian War in the nineteenth century. The protracted conflict, followed in American periodicals, ended in 1864 with the Ottoman Empire offering refuge to Circassians forced to emigrate by the Russian victors.

As Cristanne Miller has pointed out, the history of the Circassians poses all kinds of access points for Dickinson to weigh in on national and international politics, as well as a window into Dickinson's engagement with U.S. orientalism. Like Santo Domingo, Circassia could function prismatically as a lens for U.S. politics. "The image of Circassians as fiercely committed to national independence prevailed," Miller notes. "To be the child of a 'Circassian Land' was to belong to a besieged Muslim people celebrated as heroes, mythologized as exceedingly beautiful, and associated with slavery in Turkish harems."[17] One could easily imagine Circassia as a means by which Dickinson thought through antebellum racial politics. Yet if reading the poem for exclusively aesthetic purposes renders it relatively inert, reading it for solely geopolitical ends can obscure just as much once the materiality of the flowers—and the spring they conjure—ceases to perform as anything more than metaphor. Instead we might look at the way the poem connects plant life that annually "resumes" in place and the transport that local plants might inspire.

In this sense the poem performs—like the conservatory or the seed catalogue—an act of geographical compression that troubles the relationship of foreground and background and registers the potentially disorienting effect of international plant circulation. By gesturing to a foreign historical context of local flowers, Dickinson sustains a relationship among local aesthetics, local materialism, and contestation over faraway land. The multiple questions in the poem's final stanza render the local environment legible not as a fixed entity but as a confluence of migratory forces. The flowers growing locally are a reminder to the speaker that the surrounding environment is socially and politically constructed. Nor is it merely the backdrop against which political events are worked out. Depicting an active landscape defies the position of writers and thinkers from Jefferson and Crevecoeur onward who sought cultural specificity in the particular biotic qualities of a regional or national geographic range.[18]

It also resists the neat rhetoric of cultivation and human control. The commander-less nature of the flowers in "Some Rainbow" might be their most provocative element, for beyond tying the floral to the political, the language of absent leadership challenges a simple narrative of human agency. Darwin's watershed release of *On the Origin of Species* in 1859 helped popularize the concept of autonomous plant mobility, and while the impact of Darwin's text on Dickinson's social milieu has received a fair amount of critical attention, the impact of his theories of geographical mobility deserves a closer look.[19] Darwin believed that plants opportunistically dispersed their seeds. While mammals have not been able to migrate with as great a range, he notes, "some plants, from their varied means of dispersal, have migrated across the vast and broken interspace."[20] Darwin himself experimented with submerging seeds in water and testing their potential to germinate after a fixed amount of time, noting that dried seeds might travel over nine hundred miles in sea water and then sprout. Alternatively, they might travel across large distances in driftwood, in bird carcasses, in fish, in the beaks and feet of living birds, or buried in icebergs. Darwin's position on botanical migration in *Origin* is by and large focused on forces outside human control that drive evolution and change. Thoreau takes up a similar approach in "The Succession of Forest Trees," concluding that pine and oak forests often replace one another when a seed "is transported from where it grows to where it is planted . . . chiefly by the agency of the wind, water, and animals."[21] Such theorizing about environmental agency dealt a blow, as Gillian Beer notes, to anthropocentric narratives of the world.[22]

A number of Dickinson's poems emphasize botanical mobility or the mobility of pollinators like bees and butterflies. In presenting the motion of natural phenomenon, Dickinson provides an alternative narrative to the kinds of botanical circulation fostered in imperial contexts. In "The Wind Didn't Come from the Orchard—Today" (F494) the wind, "a transitive fellow," carries a burr to the doorstep, leaving the occupants inside to wonder how far the seed has traveled. In "As if Some Little

Artic Flower" (F177 A), the flower moves "down the Latitudes" in the first part of the poem:

> As if some little Arctic flower
> Opon the polar hem
> Went wandering down the Latitudes
> Until it puzzled came
> To continents of summer—
> To firmaments of sun—
> To strange, bright crowds of flowers—
> And birds, of foreign tongue!

The flower's displacement here is the central subject of the poem, though its movement emphasizes its own agency in the process: its procession down the latitudes by wandering makes it fully in charge of its own motion. Examining ideas of causality in environmental history, Timothy Mitchell has noted how "as one unravels these interwoven forces, human agency appears less as a calculating intelligence directing social outcomes and more as the product of a series of alliances in which the human element is never wholly in control."[23] While Mitchell's criticism emerges out of a contemporary posthumanist turn, the scientific idea of biological mobility was in popular circulation since at least the release of *On the Origin of Species* and is certainly at play in Dickinson's poetry.

Dickinson's access to plants also included those acquired through noncommercial routes, such as those that had crossed thousands of miles enclosed in the letters of overseas friends and correspondents. Most notably, Dickinson received a number of plant specimens from Southern Europe, the Middle East, and Asia.[24] Amherst College prepared many graduates for missionary life, and it is probable that these cuttings came from one or several of her friends who married Protestant missionaries and subsequently traveled abroad.[25] One likely candidate is Abby Wood Bliss, who moved to Syria with Daniel Bliss shortly after their marriage in 1855. Dickinson's collection of labeled plants from foreign climes includes pressings from India, Germany, Italy, Palestine, Greece, and Lebanon. Thus flowers for Dickinson might just as readily conjure international correspondence with a dear friend as the garden

Figure 3.1. Flower from "Desert of the Dead Sea" from Dickinson's botanical collections. By permission of the Houghton Library, Harvard University, MS Am 1118.13.

corners and greenhouses she knew so well.[26] Such friendship networks, as Jim Endersby has noted, constituted an important dimension of the conduct of natural history.[27] We know that epistolary correspondence shaped ideas about what social relations might mean in a world of increasing geographical mobility, but far less attention has been paid to the fact that these letters could, and often did, contain cuttings from thousands of miles away.[28] Whereas letters are usually read foremost for their textual qualities, the inclusion of plant matter asserts the materiality of the landscape from which it is removed. Dickinson cryptically titled one of her foreign pressings from a friend "that land," signaling an unspecified distant geography. The pressed plant itself makes "that land" at a distance suddenly tangible, palpable.

The epistolary exchange of botanical matter maps these informal networks onto national and religious cartographies.[29] In "Between My Country—and the Others" (F829 A) Dickinson makes flowers central to the act of communication across a great distance:

> Between My Country—and the Others—
> There is a Sea—
> But Flowers—negotiate between us—
> As Ministry.

The gulf created by the first two lines, the separating sea, is navigated—or rather importantly, negotiated—by flowers. And notably, despite the emphasis on difference of place, accentuated by the dashes, the poem makes no distinction between political, social, or religious ministry, or between personal correspondence and public federation. The multiple meanings of ministry in the final line speak to Dickinson's acknowledgment of the overlapping registers affixed to floral language. Moreover, given Dickinson's participation in a culture of sending pressed flowers in letters to friends, the poem links the sentimental circulation of flowers abroad to these other kinds of ministration.

To see that Dickinson's nature is dynamic in geopolitically significant ways is to dislodge the regionalism often associated with domestic flora. Plant circulation eroded singularity of place. The popularization of conservatories, like the one attached to Dickinson's house, meant that plants from warmer climates could survive bitter New England winters.

As Dickinson writes in an 1856 letter, "My flowers are near and foreign, and I have but to cross the floor to stand in the Spice Isles" (L315).[30] Such compression, and the ability to create a synecdoche of the world's greenery at home, revises the idea of nature in place. It makes the designations of near and far harder to distinguish and, in doing so, establishes correspondence between local and international events. It also presents a wide-angle lens on agency. We may be familiar with nature as an omnipotent and aleatory force in Dickinson's poems—the sudden storm, volcanic eruption, or "imperial Thunderbolt"—but plants are similarly surprising for their unique ontology and ability to act in autonomous, organized, and collaborative ways.

Plant Sensibility

If Dickinson's understanding of plant life challenges us to consider plant and human behavior on a global scale, her poems also encourage us to think expansively about a sensate environment. That is, Dickinson's interest in plants as organized and autonomous entities went beyond their ability to move and extended to the more radical possibility of their capacity to feel. To the extent that Dickinson anthropomorphized flora, as in flowers that "put their nightgowns on" (F127 A), she also wrote poems that acknowledged what Theresa Kelley calls plants' "material strangeness."[31] Nineteenth-century students—as well as botanists and writers—had to contend with plants as substantive objects as well as symbols.[32] And this in turn presented interpretive challenges for young botanists because the idea of plants as collectible objects dovetailed with the idea that plants could be vital subjects.

Floral vitality may seem a surprising interest for someone whose passion for plants is best archived in the herbarium of dried flowers she created as a student at Amherst Academy. The botanical instruction Dickinson received at an early age presented the natural world in a manner befitting the linked imperatives of domestic and colonial order. Her exquisite herbarium, for instance, reflects the taxonomic system used to control and regulate an increasingly networked array of natures across the globe. Identifying plants according to a taxonomic system promised to give students authority over a passive vegetable kingdom, and Dickinson's herbarium reveals the control she exerted in arranging pressed specimens from various locales. She included not only those flowers found in the wild, like marsh bellflower and frostweed, but also those growing in the garden, like privet, a common border shrub, and Persian lilac.[33] The herbarium reflects Dickinson's own choices, orienting our thinking toward plants as the recipients of human values.

Students of botany in mid-nineteenth-century America were also immersed in figurative associations between flowers and human society that likewise made plants symbols of human thought and feeling. Linnaeus's sexual basis for his classification system supported comparisons between courtship and floral pollination.[34] James Guthrie has argued that many of Dickinson's poems about bees and other floral pollinators can be read in this vein.[35] The language of flowers was another important, often related, figurative context. Elizabeth Petrino and Judith Farr have thoroughly illustrated how Dickinson often drew on this symbolic register, in which each flower species represents a specific human emotion or virtue.[36] Yet for all that Dickinson engaged with these traditions, her interest in plant life at times challenged the anthropocentric flattening inherent in both scientific and sentimental taxonomies.

By Dickinson's lifetime, plant sentience was an established, if hotly debated, current of thought within natural history. A brief discussion of this tradition is helpful for recovering structures of meaning that informed Dickinson's own engagement with plant life. By the end of the eighteenth century, naturalists hypothesized that plants could feel in ways analogous to human feeling. Whereas Linnaeus distinguished plants from animals in *Philosophia Botanica* (1751) on the very basis that plants had "growth and life" but no "feeling," by the time American naturalist William Bartram published his *Travels* in 1791, his observations led him to conclude that "vegetable beings are endued with some sensible faculties or attributes, similar to those that dignify animal nature."[37] Calling *Dionaea muscipula* (Venus flytrap) "sportive" and noting the "artifice" with which they "intrap incautious deluded insects," Bartram perceived plants as like humans in certain respects and indeed compares climbing vines to "the fingers of a human hand."[38] As Michael Gaudio notes, such a doctrine of plant feeling made sense in an Enlightenment context in which the natural and the social "were understood to operate according to the same principles."[39]

By the early decades of the nineteenth century scientists increasingly strove to detach subjectivity from empirical analysis, positioning the natural and the social further apart. Discussions of plant feeling among nineteenth-century botanists reveal ambivalence about the affinities between plants and humans. Many were willing to conceded plant sentience, but diverged widely on how to understand it. Georges Cuvier clearly limited the likeness between plant and animal, whereas Darwin's research into plant motion led him to conclude that the root of a plant "acts like the brain of one of the lower animals."[40] Augustin de Candolle, who dropped acid onto plant leaves to test their irritability and responsiveness, did not grant plants a close relationship to humans but declared that "plants live, not merely in the common sense of the word, which includes activity of every kind, but in that stricter sense, by which a higher and self-dependent activity is expressed."[41] And Harvard botanist Asa Gray, whose work Dickinson knew, vacillated over several decades, but edged toward a theory of plant intelligence, titling his 1872 children's botany textbook *How Plants Behave: How They Move, Climb, Employ Insects to Work for Them, &c.* As one 1873 review of the book noted, Gray's language "goes far toward warranting the opinion that plants are sentient creatures. If this be so, what a world of strange revelations awaits some fortunate investigator! He—or the boon may fall to the lot of a woman—will tell us if it be true that plants have pleasures and pains, that they weep when bruised, that they sleep at night, that, like the *Vallisneria spiralis*, all flowers love. . . . We might say that in the very title of his book Prof. Gray concedes the sensibility of plants, and half admits their intelligence."[42] Beyond granting female botanical authority, the reviewer reveals the extent to which this book potentially opens the door to a new paradigm of thought about plant life, one granted authority through scientific investigation. "To behave," so the reviewer crucially continues, "implies a knowledge of propriety; and if plants approach humanity so closely, it would seem absurd to deny their near relationship." In other words, plants may act in ways that are volitional, even decorous, meaning that the scientific observation of plants might call for more than simply recasting preexisting scientific categories. A "near relationship" between plants and humans potentially demands an affinity that is at once physiological, cultural, and aesthetic.

A number of popular journals helped disseminate the idea to a broader audience. An 1863 article in the *Horticulturalist* asks bluntly, "Is the plant stupid?" The author extols plant intelligence: "Who will now undertake to say that a plant is not sensible? If you go into the fields, you will tread upon a multitude of flowers that know better than you do which way the wind blows, what o'clock it is, and what is to be thought about the weather."[43] In 1870 the *Independent* reprinted an article from Charles Dickens's *All the Year Round* titled "Have Plants Intelligence?" that begins by asking, "If the oyster fastened on the rock can feel, why not the rose or the convolvulus, or the great oak tree that is fast rooted in the ground?"[44] And an 1873 article in the *Youth's Companion* on plant sleep notes that "the deeper we search into the mysteries of vegetable life, the closer appears its relations to animal existence. Botanists—especially among the French—assert that plants breathe, work, sleep, are sensitive and capable of movement. These points lead to debatable ground."[45] This "debatable ground" is the extent to which commonality exists between plants and animals and was contested because its implications were so potentially explosive. To follow the debate to its most radical social conclusions—a point before which most writers stopped—might be to concede plant life is intelligently organized in a manner that might entail ethical and epistemological demands on humans.

In making the popular argument that readers should consider the idea of plant intelligence, many articles pointed to evidence in literary and religious texts dating back centuries. The author of "Have Plants Intelligence?" points to Ovid and Hafiz as well as Roman mythology and Greek superstition to counter Linnaeus's claim that plants lack feeling, observing that science's "elder sister, Poetry, often sees further and deeper into things than [science] does." Another article on "The Soul of Plants" for *Appleton's Journal* cites Anaxagoras, Pythagoras, and Plato among those who believed in the existence of "an intelligent principle, or soul, in plants."[46] To support this idea the author describes a number of experiments on plant sensitivity, reaching back to Pliny the naturalist's observation of a tree whose leaves drooped when touched. In reaching back into the past to support contemporary ideas about plant intelligence, both articles suggest the extent to which contemporary science might look to poetry for insight into the mysteries of plant life.

The fraught nature of the debate over plant sentience in part hinges on the ways in which plants model life itself differently. Competing theories about the nature of life were widespread in the nineteenth century, offering a somewhat chaotic and contradictory range of ideas about where liveliness originates and how it is organized. Benjamin Rush believed in what Monique Allewaert has termed "vitalist materialism": a theory that matter has a capacity for life but requires external stimuli.[47] For Coleridge, life occurs when a supernatural force—a kind of spiritual antecedent—animates nature.[48] Thoreau condemns the physiologist "in too much haste to explain [plant] growth according to mechanical laws" and muses that "the mystery of the life of plants is kindred with that of our own lives." He urges a kind of restraint toward the question of life itself, arguing that "we must not presume to probe with our fingers the sanctuary of any life, whether animal or vegetable; if we do we shall discover nothing but surface still, or all fruits will be apples of the Dead Sea, full of dust and ashes."[49]

As discussion of plant vitality broadened, the distance between literary and scientific ideas about plant life narrowed. An 1878 article in the *Eclectic Magazine of Foreign Literature* makes the point about plant perception more poetically and definitively. Wordsworth's "belief . . . 'that every flower / Enjoys the air it breathes'" had in fact been validated by "the rapid march of investigation within recent years [that] transformed a poetic thought into a dictum of natural science."[50] Experiments with food, locomotion, and sensitivity had pointed biologists toward the overwhelming conclusion that the essential characteristics of plants and animals could not be easily or confidently distinguished. Part of the confusion stems from the creation of categories, for if differences might be clearly discernable "between the higher animals and plants," still "any definition of an animal or of a plant, to be either satisfactory or useful to the scientific man or to mankind at large, must include all animals and all plants."[51] Such broad categories produce strange bedfellows, as Harriet Ritvo has noted with animal classification, and point to the ways in which botanical research in the second half of the nineteenth century did not simply seek to reify Linnaean order but rather continued to revise the way in which life was conceptualized, organized, and understood.[52]

Beyond her access to a number of prominent periodicals, Dickinson directly encountered theories of plant sentience in a number of ways. The botany that Dickinson learned as a schoolgirl was mainly taxonomic, but even botanical educators who sought to teach students easy means of classifying nature conceded that the line between plants and animals was at least somewhat blurry. Almira Lincoln Phelps, the author of the most popular botanical textbook in nineteenth-century America, acknowledged the "almost imperceptible gradations by which the animal and vegetable kingdoms are blended." Toward the end of *Familiar Lectures on Botany*, a textbook Dickinson used, Phelps furthermore attributed sensation and instinct to plants even as she qualified their "principle of life" relative to that of animals.[53] The geologist Edward Hitchcock, who taught natural sciences at Amherst College and whose book on local flora Dickinson also used, likewise believed the subject to be important enough to raise in his introductory lecture on botany, noting that "the lowest tribe of animals comes nearest to the lowest order of plants. This destroys the idea of a regular chain."[54]

Prompted by her friend and distant cousin William, Dickinson also read X. B. Saintine's popular novel *Picciola* (1836), which recounts a prisoner's obsession with the plant growing in his prison courtyard—an obsession that takes the form of compassionate care and empirical analysis of the plant. The prisoner's redemption lies in his studious love of the plant, which he calls his "benefactress," and with which he comes to feel a "mysterious sympathy of nature."[55] In a letter to her cousin thanking him for sending her the book, Dickinson conflates book with plant: "Tis the first living thing that has beguiled my solitude, & I take strange delight in its society" (L27). Dickinson's first-person voice here echoes the perspective of Saintine's protagonist, whose social world for several years is primarily a flower that captivates him with its "sensibility and action."[56]

It is in light of this extensive discourse about plant feeling, as well as Dickinson's passion for gardening, that we might better hear Dickinson's own imaginative engagement with the vitality of plant life. While both botanical classification and the language of flowers were widely used in poetic figuration across the nineteenth century, a poem like "Bloom—Is Result—to Meet a Flower" (F1038 A) encourages us to consider the natural world beyond its human analogy. The poem in fact closely tracks the strategies plants use to survive through maturity:

Bloom—is Result—to meet a Flower
And casually glance
Would cause one scarcely to suspect
The minor Circumstance

Assisting in the Bright Affair
So intricately done
Then offered as a Butterfly
To the Meridian—

To pack the Bud—oppose the Worm—
Obtain it's right of Dew—
Adjust the Heat—elude the Wind—
Escape the prowling Bee—

Great Nature not to disappoint
Awaiting Her that Day—
To be a Flower, is profound
Responsibility—

When approached through conventional associations with sentimental flora, the poem proffers an analogy in which the growth of the flower stands in for the socialization of the human subject. In this vein, the flower's "Responsibility" is proxy for human subjectivity and experience, as Judith Farr suggests: "Just as the world of flowers represents the world of men and women, so certain flowers represent specific qualities or endeavors, functions, or careers."[57]

Yet the subject of the poem encourages us to think of the flower also as a living entity in its own right, one whose difference from the human makes it captivating and whose workings increasingly appeared to scientists as a self-determining power bordering on agency. The poem charts an effort to understand plant life as dynamic. Bloom might be the result that allows us to meet a flower, but the poem is less interested in the result than the process, moving backward from the presentation of the bloom through its journey of coming into flower. To survive, the flower must nearly simultaneously "pack," "oppose," "obtain," "adjust," "elude," and "escape." Rather than a static object, it is an agent in a process, and the poem as a whole draws from nineteenth-century scientific discussions of plants as adaptive—and adept—agents.

To engage the matter of plant liveliness, Dickinson's approach to organic matter combines scientific knowledge with a sentimentalism often considered antithetical to empirical study. For Dickinson, shared sentiment might become a way to connect the human condition to that of birds, flowers, and the natural environment at large. In "The Birds Reported from the South—" (F780 A, 1863) this rapport becomes increasingly intimate:

The Birds reported from the South—
A News express to Me—
A spicy Charge, My little Posts—
But I am deaf—Today—

The Flowers—appealed—a timid Throng—
I reinforced the Door—
Go blossom to the Bees—I said—
And trouble Me—no More—

The Summer Grace, for notice strove—
Remote—Her best Array—
The Heart—to stimulate the Eye
Refused too utterly—

At length, a Mourner, like Myself,
She drew away austere—
Her frosts to ponder—then it was
I recollected Her—

She suffered Me, for I had mourned—
I offered Her no word—
My Witness—was the Crape I bore—
Her—Witness—was Her Dead—

Thenceforward—We—together dwelt—
She—never questioned Me—
Nor I—Herself—
Our Contract
A Wiser Sympathy

The poem refuses a firm distinction between the civic and the natural through language that overlays the conflict of the Civil War—the reports from the South, the "News," the "spicy Charge"—with the progression of the seasons.[58] Not wholly distinct from human affairs, nature is also no mere mirror of the speaker's inner mournful state, but "A Mourner, like myself." Nature is a sensible agent that strives for notice, draws away austerely, and ponders her own frosts. We could read this as simple anthropomorphism or the pathetic fallacy, connecting the outer world to the speaker's inner turmoil, except for the fact that the poem tracks the discord between the speaker and nature before "I recollected Her—," making nature's emotions autonomous. Furthermore, Dickinson relies on metonymy to destabilize the trope of "nature." Nature includes the birds with their spicy charge, the throng of flowers, the frost; rather than a steady entity it is an assemblage of interactive parts moving in time and space.[59]

Yet if nature does not simply reflect the speaker's inner grief, it also does not stand apart completely from that grief either. The relationship culminates in the poem's last stanza where the speaker and the natural world are bound by a "Wiser Sympathy" that forms "Our Contract." This closure by contract makes wordless, essentially private states of grief into something shared. The private nature of affect, as Lauren Berlant notes, can close off the possibility of addressing pain and suffering through political channels.[60] Here, pain is private, but its very intimacy is the basis for contract, and its signs are public: the speaker's "Crape" and nature's "Dead" are both witness to their suffering and visible markers of feeling. The terms of the contract produce an ecological sensibility that makes affect an imperfectly shared trait.

By resisting the metaphorical, this sentimental connection challenges the notion that feeling is essentially a human characteristic. Bodily sensation is critical to aesthetic theories in the eighteenth and nineteenth centuries, and the sentimentality epitomized in the United States by writers like

Susan Warner and Harriet Beecher Stowe was indebted to the eighteenth-century discourse of sensibility.[61] While the political dimension of this tradition—its role in defining, to quote Elizabeth Maddock Dillon, "the terrain of liberal subjectivity"—is well established, scholarship on the link between feeling and aesthetic subjectivity has focused on an exclusively human politics.[62] Dickinson's writing points toward a different kind of sentimentalism in that the kind of subjectivity that she attributes to the natural world across her poems and her letters blurs precisely the borders that distinguish plants, animals, and humans. The kind of sensory experience that Dickinson attributes to plants and animals smudges to the point of erasure the separation of natural object from perceiving subject.[63]

By refusing the rigid separation of human, animal, and plant, the poem unsettles a fundamental premise that shaped nineteenth-century conceptions of personhood. That the speaker shares a sympathetic contract not with other humans but with nature invites identification across category distinctions. How the line gets drawn between sentience and sensibility, between raw physiological feeling and discernment, mattered in nineteenth-century America because taste had such political purchase. In William Huntting Howell's words, "Understanding sensibility promised to answer the question of what counts as personhood." But what if sensibility did not only "index humanity" per se?[64] Dickinson's perspective on flora blurs the boundary between human and plant by at times making sentiment a character trait that she shares with the natural world, rather than merely an aesthetic stance toward that world.

In "Flowers—Well—if Anybody" (F95 B) the implications of a sensate environment extend to aesthetic judgment. The poem moves from a human experience of feeling to that of butterflies through their shared appreciation of the botanical:

> Flowers—Well—if anybody
> Can the extasy define—
> Half a transport—half a trouble—
> With which flowers humble men:
> Anybody find the fountain
> From which floods so contra flow—
> I will give him all the Daisies
> Which upon the hillside blow.

Too much pathos in their faces
For a simple breast like mine—
Butterflies from St Domingo
Cruising round the purple line
Have a system of aesthetics—
Far superior to mine.

The first stanza reads as a rehearsal of the sublime: the mix of pleasure and pain, the ineffability of the source of such overwhelming sensation, and the dwarfing power of the object to humble men. One could read it as a constitution of the self through aesthetic appreciation of the natural world. In this sense, the flowers incite "extasy" and evoke both "transport" and "trouble" in the perceiving human subject. At the same time, the "extasy" of the flowers makes the nature of the interaction more ambiguous. The flowers quite possibly "humble men" because they demonstrate, rather than inspire, an ecstatic state. The ambiguous syntax emphasizes the ambiguous dynamic, or at least the difficulty of defining the nature of the interaction in language. The playful proposition of Daisies for a definition underscores the meaningfulness of this equivocation. If the power of the flowers could be articulated, then the speaker could control their circulation. The hyperbole of offering "all the Daisies" upon the hillside (which carries the attendant and ironic presumption that they are property to be gifted) suggests the unlikeliness of such a causal definition.

The second stanza of the poem continues to challenge the subject/object orientation of a sublimely constituted subjectivity as well as a hierarchy that runs from humans down to animals and then plants. In the lines "Too much pathos in their faces / For a simple breast like mine—" the flowers are given "faces" whereas the speaker synecdochically becomes a breast. And here again the poem prevaricates about where the pathos resides: in the eye of the beholder, or in the flowers themselves. Furthermore, the excess here is beyond the speaker's ability to feel, but not the butterfly's. The rest of the poem continues to trouble the exclusivity of affect to human subjects, and flowers serve as a bridge between the "I" in the poem and the butterflies. The final lines disturb the trajectory of sublime revelation by the observant speaker and attribute the notion of aesthetic evaluation to the butterflies. That it is not simply a transposition of the speaker's own experience is clear from the comparative aspect of the final lines: this system of aesthetics is "Far superior to mine."

To see butterflies as having a system of aesthetics is to strip away the notion that aesthetic judgment is exclusively a human experience. And in acknowledging the sensate dimension of the butterflies' existence, the speaker describes a world in which the nonhuman performs a process of feeling and judging that is usually only consigned to "persons." Here the attraction of the butterflies to the beauty of the flowers edges toward animal aesthetics. This sensibility in turn has political implications, for it suggests that nature "federates" (F798 A). Historians and literary critics have demonstrated at length how sensibility was a central concept for social transformation in the revolutionary period of the late eighteenth century.[65] At least one of those revolutions seems to be invoked in "Flowers—Well—if Anybody": that the butterflies hail from Santo Domingo evokes the specter of the Haitian Revolution. As Ed Folsom and Kenneth Price have argued, San Domingo was a word that by the mid-nineteenth century had "become inscribed in the language in a kind of shorthand" for the Haitian slave revolt and subsequent revolution.[66] In Dickinson's poem, the sensibility that is brought into focus from Santo Domingo does not separate a new social order from the old, but rather casts into chaos the very classificatory foundations of racial and ecological hierarchy. The butterflies' aesthetic ability to discern, presumably, which flowers to select for nectar is deemed "superior" to the speaker's subjectivity constituted through the sublime experience of the flowers. This flipping of the script renders the boundary between human and nature, or figure and ground, indeterminate.

An aesthetic that corresponds with a discerning, sensate environment challenges the correlation of rights with the individual affective citizen-subject. A feeling nature is, aesthetically speaking, potently political, challenging individualism not as a matter of law or even race but from a biological perspective. The sensibility that Dickinson describes in the poem augurs a different kind of revolution in the way that the relationship between environment and politics is imagined and articulated. By identifying other kinds of biological life as having characteristics foundational to the way the citizen-subject is recognized, Dickinson undermines claims to human exceptionalism.

Other Dickinson poems redraw the boundaries of sensibility to extend beyond conventional notions of human polity. At times this takes the familiar form of personification, as in "Nature Affects to Be Sedate" (Fr 1176), which treats Nature as a "spacious Citizen," reconceptualizing both nature and citizenship:

> Nature affects to be sedate
> Opon Occasion, grand
> But let our observation shut
> Her practices extend
> To Necromancy and the Trades
> Remote to understand
> Behold our spacious Citizen
> Unto a Juggler turned—

The poem begins by emphasizing the limits of "our observation" to understand. Nature puts on appearances that are performative rather than essentialist, and the speaker suggests that empiricism may not yield full understanding of "Her practices." Instead the poem suggests we "let our observation shut" as nature's practices extend to what is remote to our understanding. Nature here asks for a different form of attention.

Yet what is most puzzling is not that nature is mysterious and remote, but that nature is "Citizen": a political subject. Personified nature is a common enough trope, but here "spacious" as a modifier disrupts the idea of nature—or nation—as a bounded, knowable body. Moreover, characterized as a "Juggler," Nature turns into an active performer who showcases a dynamic process.

And for all that Dickinson was drawn to the ways natural entities could unsettle human modes of organization, the poet's feeling nature at times checks claims to human superiority by signaling to modes of organization beyond human comprehension. Poems like "The Veins of Other Flowers" (F 798) illustrate the complexity and sophistication of plant life and ecological processes:

> The Veins of other Flowers
> The Scarlet Flowers are
> Till Nature leisure has for Terms
> As "Branch," and "Jugular."
>
> We pass, and she abides.
> We conjugate Her Skill
> While She creates and federates
> Without a syllable—

On a formal level, the poem works to mitigate an anthropocentric politics that Dickinson posits is inherent to specialized plant terminology. The opening lines already pose a contradiction. "The Veins of other Flowers / The Scarlet Flowers are" is both metaphor and metonym when the scarlet flowers are turned into veins by metaphor and framed as parts of a larger whole (the other flowers) through synecdoche. Such a paradox of simultaneous substitution and contiguity drops the scarlet flowers from federation within language, "Till Nature leisure has for terms / As 'Branch' and 'Jugular.'" Nature has no leisure for such terms, as the pace of the poem implies, and organizes according to a logic that cannot be reduced to human models. The terms "We conjugate," Dickinson suggests here, work to impose an artificial order—a political order, as suggested by "federates"—that runs counter to nature's wordless organization. Language cannot fully convey natural processes because Nature, for Dickinson, is not organized like a language. The imposition of terminology flattens comprehension and imposes an imagined structure of relation where parts service the whole. "Branch" and "Jugular" are, like "syllable," a discrete part of a larger whole. The rejection of language is also a rejection of the logic of language, which under grammatical rules depends on parts to constitute a whole, and a whole to understand parts. The poem resists this mode of relation, suggesting that nature's operations are not so easily patterned.

* * *

Dickinson's poems and letters help us recover an important facet of the way that nineteenth-century scientists, poets, and readers understood plants. Not simply passive objects receptive to human needs and desires, they behaved in ways that challenged assumptions about stasis and even rootedness. Whereas political efforts to document, categorize, and harness useful plants had long shaped their representation in pedagogical materials, writers could turn to theories of plant vitality and adaptation to unsettle associations with stability and essentialism.

While the idea of continental biotic distinctiveness served the ideology of American exceptionalism, Dickinson encountered a landscape changed not only by local industry but by the burgeoning marketplace in seeds and plants from around the globe. The powerful global market in plants transformed the practice of cultivation—introducing new plants and new technologies. If nineteenth-century home gardening has thus far largely been perceived as a fairly parochial activity, Dickinson's botanical poems demonstrate our need to think anew about what it means, botanically, to be at home.

So too do the actions of plants themselves, whether propagating of their own accord or failing to thrive. The multispecies turn in anthropology and the nonhuman turn in literary criticism have taken up the political and ethical question of how we incorporate other living entities into our worldmaking, although they have largely focused on animals.[67] Plant studies is a small but growing field that helps unsettle assumptions and make sense of the epistemologies and habits of mind that limit our comprehension of other forms of life. Dickinson's fascination with plants reminds us that they constituted an important and wonderfully difficult category of scientific thought in the nineteenth century. And her writing illustrates the power of poetic experimentation in imagining life more capaciously. If Dickinson shows us the ways that plants pushed at the limits of language, several of her contemporaries sought to apply such lessons to the social realm. As the next chapter shows, one line of abolitionist argument lay in emphasizing the importance of horticultural alternatives to the plantation system.

4

Botanical Abolitionism

In August 1855 a short article titled "Meditations from Our Garden Seat" appeared in the *Independent*, a New York–based weekly paper with a Congregationalist focus and an antislavery stance. The author opens with a reflection on the responsibility of the gardener: "The Great Father did not give man and woman a garden ready tilled—nor endow them with an angelic gardener whose superior care should do all the thinking and laboring for them, [but] a garden to till and dress for themselves." Though the piece makes no explicit mention of slavery, it applies this biblical observation to an interpretation of labor and ownership that is based not on laws and documents but on the act of cultivation. Dismissing money as a basis for possession, the article stresses that "he alone owns a garden, in any true sense of the word, who plans and cultivates himself—whose own labor and thought entered into the growth of it. The jaded millionaire . . . commissions some poor Smith or Jones to lay out and plant and tend one for him, [but] he is not in any blessed sense the owner of it." To own a garden in this sense is an act of self-cultivation as much as tending to the soil; it means caring for land and self in tandem, thereby transforming both. To the person who works with plants, "every plant has its history, and a part of his own history, a portion of his own life, has grown up into it, and with it." Gardening "seems to bring a man into confidential relations with all the forces of nature. A man comes to have in himself a plant-life, a plant appreciation of sun, rain, wind, and all the mysterious agents of natural life."[1] Plant and gardener are entangled here, evident even in the syntax. For this meditator, garden ownership lies, in a Lockean sense, with the laborer whose own life merges with that of the plants in his care. The author appeals to readers to reflect on the value of gardening, but also to look out on civic society—and its most intractable issue—from the perspective of the garden seat.[2] In this vein, raising plants might subtly teach a horticultural nation how to adopt an antislavery orientation.

The meditator was Harriet Beecher Stowe, who had published *Uncle Tom's Cabin* three years prior, and who by 1855 was one of the most famous white members of the abolitionist movement. She was also a passionate gardener with an interest in the civic as well as religious lessons to be gleaned from the garden. The benefits of gardening for Stowe were multiple, as she rehearses across "Meditations," but none more so than the way that caring for plants encouraged a sympathetic orientation to nature as a whole. As we saw in the last chapter, many scientists and gardeners in nineteenth-century America grappled with the political implications of plant vitality. Stowe drew on the garden as a tool in her abolitionist work. The same year that she published "Meditations from Our Garden Seat," Stowe began working on *Dred: A Tale of the Great Dismal Swamp*.

Dred is an abolitionist text centered around a series of interrelated domestic plots, but it is a very different book than the novel that made Stowe famous, representing a shift in strategy that reflected both the changing political climate and her own evolution in thought. In *Dred* Stowe presents the possibility of violent revolution before ultimately suppressing that idea in favor of modeling interracial community. In order to make sense of the novel's contradictory impulses, scholars have emphasized its departure from a politics of feeling toward a focus on legal and political channels that might produce structural change.[3] But Stowe's second abolitionist novel is also significant for the way that it represents the natural world, and especially its botanical profusion, as sympathetic to abolitionist principles.[4]

Her novel presents a world teeming with growth and vitality that is at once personal, environmental, and institutional. Tapping into the country's botanical zeitgeist, Stowe sought to show how this growth could best be managed outside the plantation model by aligned national progress with the kinds of sentimental cultivation found in the domestic sphere. For Stowe, as for many of her readers, horticulture constituted a vital form of civic engagement. Slavery depended on strict regimes of classification linked to equally strict practices of cultivation that sustained plantations as economic units. But in the middle of the nineteenth century, new theories of plant growth and vitality showed that in fact plants could neither be classified nor cultivated so neatly.

Applying these ideas in *Dred*, Stowe models a world in which plant life thrives outside the plantation structure. Her characterization of plants reframes our understanding of what might be called sentimental science and shows how domestic cultivation practices were seen as central to the practice of democracy in the decade preceding the Civil War. Sentimental cultivation would promote an understanding of the natural world that collapsed the hard boundaries between home and wilderness, Black and white, humans and plants. By resisting rigid scientific categories and binary thinking, Stowe instead emphasized that the crossing, hybridity, and prodigality familiar to the domestic naturalist could form the basis for a different, more egalitarian political order. *Dred* represents a sustained attempt to translate the concept of botanical growth into lessons for a country in crisis.

A Gardener's Eye to Order

Stowe loved gardening. As a young wife in Walnut Hills, Ohio, Stowe planned elaborate gardens and spent countless hours working on them. Susan Monroe Stowe wrote of her mother-in-law, "Mrs. Stowe [had] a passionate love of flowers, and always found time to cultivate house-plants in winter, and in summer to gather and arrange flowers for decorating her rooms."[5] These interests are confirmed in letters that Stowe wrote to her siblings, in columns she wrote for the *Independent* and the *Atlantic*, and in her husband's complaints that she spent too much money on her gardens.[6] Stowe's descriptions of plants reveal both a great familiarity with plant cultivation and a belief, shared by many of her contemporaries, in its moral and physical benefits for the cultivator.

Indeed, over the course of Stowe's lifetime botany became a household practice, recognized for its usefulness across a range of fields. The May 1836 edition of the *Family Magazine* sought to correct any notion that botany was only for doctors and surgeons, declaring that "knowledge of botany is essential to the farmer and the horticulturalist . . . the mechanic . . . the merchant . . . the apothecary . . . the clergyman [and] the lawyer."[7] This viewpoint became increasingly widespread throughout the next several decades as botanical instruction became prevalent in schools and widely practiced in the home.[8] By 1833 the subject was already being offered at both men's and women's schools, including Catharine Beecher's Hartford Seminary, where Stowe taught.[9] Some of Stowe's contemporaries set science in opposition to aesthetics, sentiment, and religion, but as Stowe believed and her writings suggested, plant cultivation was so popular precisely because it brought these domains together.

Beecher, Stowe's sister, was among a number of public figures who saw floral beauty from a philanthropic perspective, promoting flower distribution as a form of community service. Middle-class women and children should "distribute roots and seeds to those who do not have the means of procuring them," Beecher wrote, in order to "awaken a new and refining source of enjoyment in minds, that have few resources."[10] Nathaniel Bagshaw Ward, creator of the Wardian case, likewise believed that this technology for protecting plants could be used for the relief of the "*physical* and *moral* wants of densely crowded populations in large cities." As evidence, Ward cites a letter he received from a glazier who was able to create a perfect "Lilliputian landscape" despite the polluted urban atmosphere in which he lived.[11] Both Beecher and Ward sought to ameliorate the experience of poverty through plants, imagining a primarily aesthetic approach to structural inequalities.

Stowe shared her sister's widely held belief that beauty found in nature was a sign of the divine, and aesthetics remained an important aspect of her evolving thought even as she tested the power of aesthetics to transform systemic conditions. In *Sunny Memories of Foreign Lands* (1854) Stowe encapsulates the popular idea that flowers were potent symbols of social transformation because they revealed divine artistry, not simply divine logic. Struck by the wildflowers she found growing along a path while trekking in the Swiss countryside, she writes, "He who made the world is no utilitarian, no despiser of the fine arts, and no condemner of ornament." Flowers serve as a signifier of divine interest in beauty and taste, linking morality to aesthetics. At the same time, "there is a strange, unsatisfying pleasure about flowers, which, like all earthly pleasure, is akin to pain," she begins, and then asks, "In what mood of mind were they conceived by the great Artist?"[12] The commingled pleasure and pain associated with the beauty here verges on the sublime, suggesting the transformative potential of the flowers she sees. For Stowe, flowers were potent symbols of social transformation because they revealed not only the argument from design but also a kind of pleasure.

Stowe insisted on the aesthetics of divinity—a sense that beauty served a theological purpose—in order to challenge the rigors of Calvinist theology. Characters in her novels who fail to appreciate floral beauty also have difficulty sympathizing with others. In *Oldtown Folks* (1869), Stowe's study of New England life, the aptly named Miss Asphyxia appreciates only what is useful, and therefore cannot understand the value of flowers that serve no utilitarian purpose:

> [She] had one word for all flowers. She called them "blows," and they were divided in her mind, in a manner far more simple than any botanical system, into two classes; namely, blows that were good to dry, and blows that were not. Elder-blow, catnip, hoarhound, hardhack, gentian, ginseng, and various vegetable tribes, she knew well and had a great respect for; but all the other little weeds that put on obtrusive colors and flaunted in the summer breeze, without any pretensions to further usefulness, Miss Asphyxia completely ignored. It would not be describing her state to say she had a contempt for them: she simply never saw or thought of them at all. The idea of beauty as connected with any of them never entered her mind,—it did not exist there.[13]

For Stowe, Miss Asphyxia represents the failure of a Puritanism that cannot appreciate form without function, beauty for its own sake. Her harsh, practical view of the natural world is conjured by the cacophony of the vegetable names—hoarhound, hardhack. She works the ground, but she does not love it. By disregarding what is beautiful in nature, she fails to follow the divine model Stowe sees therein, and her lack of sympathetic communion with the plants around her correlates with her inability to successfully nurture Tina, the orphaned child who comes to live with her briefly before running away. For Stowe, it followed that an ability to appreciate beauty led to a sense of affective interconnectedness, whereas a purely functional perspective inhibited sentimental bonds.

Stowe's enmity toward Puritan rigor is evident throughout her writings; what is particularly interesting here is that she aligns a Puritan lack of sympathetic identification and aesthetic appreciation with the practices of botanical classification, and in doing so collapses scientific and religious imperatives to order. Miss Asphyxia's simple taxonomy limits her perception—she cannot see how individual plants fit into a connected whole, and holism was at the center of the project of botanical classification as it was popularly practiced at midcentury. Stowe renders the specific names of "blows" that Miss Asphyxia ignores, demonstrating her facility with flower identification and an intimate knowledge of the different kinds of vegetation to be found growing in the Northeast. She also points to the broader organization of flowers in antebellum society. A love of nature's beauty, for Stowe, corresponded with an understanding of its organization.

In this belief Stowe drew on the rise of botanical instruction as a domestic science that emphasized systematic thought about individual plants. In *Familiar Lectures on Botany* Almira Lincoln Phelps praises botany's systemic influence on the organization of everyday life:

> The very logical and systematic arrangement which prevails in Botanical science, has, without doubt, a tendency to induce in the mind the habit and love of order. . . . Whoever traces the system which classes the vegetable tribes through its various connexions, by a gradual progress from individual plants to general classes, until the whole vegetable world seems brought into one point of view: and then descends in the same methodical manner, from generals to particulars, must acquire a habit of arrangement, and a perception of order, which is the true practical logic.[14]

Plant science, so Phelps's logic goes here, is good for household management because it promotes a place for everything and puts everything in its place. By organizing plants systematically, the collector can gain a purview at once sweeping and particular. This ability to think expansively of the "whole vegetable world" offers a perspective that extends well beyond the domestic sphere and is not limited by geographical boundaries.[15] Stowe revered the way this kind of systematic perspective might transcend individual interest, but she challenged the inelasticity she found in both contemporary science and theology. In Stowe's late antebellum writings, this challenge takes the form of a botanical aesthetic that encourages flexible, pluralistic, and nonhierarchical thinking.

Nowhere is that aesthetic more apparent than in *Dred*, where two seemingly separate plots continually intertwine. One plot follows the eponymous Dred, a Black revolutionary who cultivates a Maroon community in the wilds of the swamp, and whose desire for racial justice is in perfect communion with the plants that grow there. The other half follows the domestic travails of Nina Gordon, the orphaned daughter of wealthy southern plantation owners who is a spoiled and materialistic but likeable coquette with a penchant for wildflowers and beautiful commodities. The two never meet, and each anchors distinctive plot possibilities for producing change; Dred makes plans for violent revolt, whereas Nina's romance plot ultimately serves as the mechanism for an exploration of potential legal and political challenges to the status quo. But despite the separation of their worlds, secondary characters continually move between their plots. Nina's half brother Harry knows Dred and encounters him often while traversing roads through the swamp. Harry declines to be involved in Dred's planned revolt because of the danger it might pose to Nina. Later, after Nina dies during a cholera outbreak on the Canema plantation, Dred saves Clayton, the man who had courted her during the first half of the novel, when Clayton is physically attacked for his antislavery sympathies.

Moreover, despite their parallel existences, Dred and Nina are connected on a figurative level by their shared identification with botanical growth. Dred is "so completely under the nursing influences of nature" as to be as "perfectly *en rapport* with them as a tree" (273–274, emphasis original). Nina, who is likened by a suitor to "a sweet-briar bush, winking and blinking, full of dew-drops, full of roses, and brisk little thorns" (304), is so familiar with the local wildlife that she "might almost have been considered one of the fraternity" (33). While Dred and Nina have different relationships to botanical language—Dred is associated with sublime Nature and Nina is more firmly linked to cultivated growth—the affinities Stowe presents between them are essential to understanding Stowe's alternative democratic vision and the power of a horticultural vernacular for her evolving sense of abolitionist strategy.

In fact, a "principle of *growth*" (496, emphasis original) structures the novel, setting up the inevitability of change and a question of how best to manage growth of all sorts—institutional, communal, personal, and botanical. This is evident from the start in the way that the narrator explains how Nina, who at first demonstrates herself to be incapable of household management, slowly develops into a model of self-sacrifice. In a familiar mode of sentimentality, the narrator depicts Nina's maturation as a kind of ethical development in communion with "the same quiet forces which swell the rosebud and guide the climbing path of the vine" (331). In this vein the narrator explains how "there are times in life when the soul, like a half-grown climbing vine, hangs wavering tremulously, stretching out its tendrils for something to ascend by" (344). Nina's trajectory follows a movement away from the kind of scientific classification Stowe associates with self-interest toward a holistic perspective associated with a communal ethic. Nina is characterized at one point as being "as full of streaks as a tulip" (304). The reference is to Samuel Johnson's *Rasselas*, whose narrator says that a poet, unlike a scientist, "does not number the streaks of the tulip."[16] Laura Dassow Walls has described an empirical holism that animates the writings of Thoreau, Whitman, and von Humboldt with a sense of interconnectivity and open-endedness. For the empirical holist, facts do not fit into a predetermined whole; rather, the whole is generated out of a constantly evolving "interaction of differences" between parts.[17] In *Dred* we see Stowe applying similar principles to the domestic narrative. Plant growth is neither predetermined nor constrained.

Framing Nina's growth toward spiritual enlightenment in botanical terms, Stowe connects Nina's development to the success of other horticultural projects in the novel.[18] One such project is the garden maintained by Tiff, an enslaved man who works for the poor Cripps family. While the family's patriarch is an abusive drunk, Tiff creates for the Cripps children the possibility for a different future. When their mother dies, Nina takes pity on the family in part because she is fond of Tiff, whose horticultural skills impress her. Tiff creates a picturesque aesthetic through both the careful manicuring of cultivated flowers and "intermingled" vines that cover the house (108). Beautiful growth is likewise celebrated at the home of Nina's enslaved half brother Harry and his wife Lisette. The narrator describes a scene of wild and cultivated plants intersperse to create a beautiful tableau: "The door, which opened onto a show of most brilliant flowers, was overlaid completely by the lamarque rose . . . and in large clusters of its creamy blossoms, and wreathes of its dark-green leaves, had been enticed in and tied to sundry nails and pegs by the small hand of the little mistress, to form an arch of flowers and roses" (55). These scenes connect gardeners across racial and class divides and stress the power of the homemaker to train growth into a beautiful scene.

Outside the mediating effects of domestic cultivators, botanical growth in the novel is a force that threatens to overwhelm, and Stowe leverages this as an abolitionist argument by showing how this is a force that animates people as well as plants. After Nina dies, her suitor Clayton continues to deepen his own antislavery stance. As Clayton tells his friend, "Minds *will grow*. They *are* growing. There's no help for it, and there's no force like the force of growth. I have seen a rock split in two by the growing of an elm-tree that wanted light and air, and would make its way up through it" (470). The narrator likewise marvels fearfully that "there is no principle so awful through all nature as the principle of *growth*. It is a mysterious and dread condition of existence, which, place it under what impediment or disadvantage you will, is constantly forcing on; and when unnatural pressure hinders it, develops in forms portentous and astonishing" (496). Starting with a force that anticipates human form, Stowe argues that it can best be managed through a sympathetic relationship with the natural world, including that embodied by Dred.

Critics have long observed how abolitionists used sympathy—a mode of identification that appealed to the ethical imperatives of the feeling subject—to push for the end of slavery. To feel right, Stowe averred in her best-selling *Uncle Tom's Cabin* (1852), was to identify with the pain of the enslaved, and to act accordingly. Appealing to the sanctity of the family and relations therein, *Uncle Tom's Cabin* begins by asking readers to feel the plight of Eliza, who must flee with her child on a dangerous journey to prevent him from being sold. Scholars have rigorously unpacked how sympathetic appeals reproduced the idea of human difference based on supposedly different capacities to feel. To feel right, as Kyla Schuller has shown, was based on a racialized and sexualized idea of the individuated subject.[19] Monique Allewaert similarly points to "the fantasy of the bounded body that may, through the working of sympathy, be motivated to care for another."[20] The sympathetic subject is an ecological construction as well as a political one, creating a sharp divide between body and environment.

Stowe's novel upholds sympathy as a model but suggests it might be not only felt between people but shared with plants, animals, and the forces of nature. Stowe depicts Dred as being most sympathetically aligned to the growth around him. As Jamie Bolker argues, in *Dred* "moral authority, sympathy with nature, and blackness are linked."[21] Dred's "rapport" with nature is like that of "a tree," which presumes interdependence, "so that the rain, the wind, and the thunder, all those forces from which human beings generally seeks shelter, seem to hold with [his body] a kind of fellowship, and to be familiar companions of existence" (273–274). Nature's "nursing" influences are here shown to be reflexive and networked. What this means, in part, is that Dred—and by extension all humans who share his "sympathy and communion with nature"—embodies nature's interdependence (274). Stowe's association of humans with an elastic botanical aesthetic ran counter to contemporary male ethnologists and their systems of racial classification, which sought proof of racial difference by projecting nature as immutable.

Stowe also depicts Dred's relationship to the natural world as falling within the sentimental tradition of her contemporaries. For example, in "Autobiographical Romance" (1840), Margaret Fuller describes how in cutting flowers from her garden, "I kissed them, I pressed them to my bosom with passionate emotions, such as I have never dared express to any human being."[22] In Elizabeth Stuart Phelps's *The Story of Avis* (1877) the protagonist has a widowed aunt who "tended her flowers as if they were all orphans, and loved that ivy like her own soul." Aunt Chloe's identification with her plants is so close that she remarks to Avis's suitor that "I can't sleep warm if I know my plants are cold. Did you never notice, Mr. Ostrander, how an arbutelon, for instance, will shiver?" If she could start life over, she says, "and choose for my own selfish pleasure" she would be "a florist, perhaps . . . or a botanist," a formulation that acknowledges both the gendering of botany and the difficulty women faced in entering scientific domains.[23] Far from being shallow allegory or simple personification, this sympathy conveyed deep knowledge. Affect, indeed, was an important part of the history of botany in this period, supporting the scientific authority of the domestic cultivator.[24]

In her novels, Stowe appealed to the dominant domestic ideology by keeping the home as the site of moral influence and careful economy, but she also repeatedly confounded the borders that were usually imposed upon the genre. As Amy Kaplan, Gillian Brown, and Lori Merish have shown, domestic ideology upheld the home as means by which identity—personal and national—was constituted. This ideology frequently upheld distinctions between "domestic" and "foreign," "home" and "abroad."[25] Home gardens were often liminal spaces, "set apart from the wildness of nature but more flexible, both spatially and socially, than the confined rooms of the home."[26] But emphasizing botanical growth as an unstoppable force, what Stowe imagines in *Dred* is a quite different vision of domesticity, where floral imagery is notably fluid and ideologically charged, moving seamlessly from parlor to garden to swamp.

A Rebellious Garden

While completing *Dred* Stowe wrote to her editor, Frederick Law Olmsted, who had recently journeyed through the American South and asked him for botanical details:

> Of what species is the Pine of which you make so great mention and of which the greater part of the Pine Forests are composed? Are the mosses and flowers which grow under them of the same species that grow in the Pine Forests in the Northern States? Did you notice that white crisp frosty-looking moss which grows on Pine lands with us?—Also the feathery green ground Pine?—Pray what is the Cat briar of which you make so frequent mention? Is the Holly like the English?—Have you ever seen it employed for hedges?—I wish very much if you are in our vicinity in Boston that you would make me a call for I should like very much to read you some parts and get a little help from you about laying out the topographical details—It is absolutely necessary for me to get a perfect definite idea of the country where I suppose the scene will be laid and in conversing with you I could do it.[27]

Such details would prove "absolutely necessary" to Stowe's depiction of the swamp.

Stowe knew that botanical enterprises shaped political praxis. *Uncle Tom's Cabin* has a much tidier plot structure than *Dred*, but already in her earlier novel Stowe identified the significance of plant diversity to American intellectual and political history. The recalcitrant enslaved girl Topsy famously evokes the relationship between plant practices and politics when she responds to an inquiry about her origins by musing, "I spect I grow'd."[28] This seemingly atheistic riposte—the correct answer to Ophelia's catechizing would have been, of course, "God made me"—makes sense in terms of Stowe's belief that organic growth was godly. Pointing to the way that slavery uprooted family ties, Topsy's response reminds the reader of the botanical practices associated with slavery.

As an economic system mainly dependent on the large-scale growth of staple crops, it starkly contrasted with the principles of ecological diversity that revolutionary era politicians perceived as a key to economic viability.[29] And while a racially diverse democracy is an idea that Stowe ultimately rejects in *Uncle Tom's Cabin* with her infamous vision of Black emigration to Liberia, she comes much closer to this vision in *Dred* when she highlights the horticultural diversity that existed outside the monolithic plantation economy.

The shift in Stowe's political vision can in part be explained by her developing botanical aesthetic, which pushes her toward a more radical vision of social inclusion. At the center of this aesthetic is the swamp with its prodigious growth. First described as "a considerable check on the otherwise absolute power of the overseer" (210), the swamp is a mutable symbol in *Dred* as much as a landscape for action: it is in one moment in "defiance to all human efforts either to penetrate or subdue" (209) and in the next "the reflection of [the mind's] internal passions" (210). Similarly, it serves at once as refuge for slaves and as allusion to the institution of slavery, "this great system of injustice, which, like a parasitic weed, had struck its growth through the whole growth of society" (391). Critics like David Miller have wrestled with Stowe's representation of the swamp, finding its symbolic resonance inconsistent and therefore a failure of the novel. But Stowe's depiction undermines the ostensible contrast between the messy swamps and refined borders and hedges of domestic space. Monique Allewaert has convincingly argued that the "liquefying natural world" of the swamp presented a broad challenge to the order of the plantation, and in linking the swamp to the domestic realm, Stowe shows that these two seemingly incommensurable spaces overlap in politically significant ways.[30]

The Great Dismal Swamp is one of the most carefully arranged spaces in the novel. Consider the following passage: "Evergreen trees, mingling freely with the deciduous children of the forest . . . afford shelter to numberless sounds. Climbing vines and parasitic plants, of untold splendor and boundless exuberance of growth, twine and interlace, and hang from the heights of the highest trees pennons of gold and purple. . . . A species of parasitic moss wreaths its abundant draperies from tree to tree, and hangs in pearly festoons" (209). Stowe's imagery here might be chalked up to an impulse toward realism, as seen in her letter to Olmsted, if it did not also resonate with the festoons and banners of flowers so prevalent in the domestic spaces she has already described. Her characterization of the parasitic moss, for instance, recalls the way Lisette drapes the lamarque rose around the cabin she and Harry share. All the growth in this passage is carefully arranged, from the "pennons of gold and purple" high in the trees to the holly's careful contrast of scarlet berry and green leaves. From the vantage point of a clearing in the swamp one might see "a grape-vine, depending in natural festoons from a sweet-gum tree, [make] a kind of arbor" (406). Contemporary newspaper and journal articles encouraged aspiring botanists to spend time "among the wild scenes of our meadows, rocks and forests,"[31] and Stowe reflects the fact that the practice of botanical collection often necessitated a fluid relationship between home and field. Stowe describes picturesque gardens naturally occurring in the wild—the mingling, climbing, twining, and interlacing of plants in the swamp that is at once organized and anarchic—and thereby affirms the mutuality of the home and the surrounding wild spaces.[32]

More significantly, Stowe emphasizes that the swamp is a domestic space for the Maroon community led by Dred. This is especially apparent when, after Cripps turns the family homestead into a grog shop, Tiff takes the children and flees into the swamp. Stowe represents it as a familiar place of comfort and moral virtue. In contrast to the bedroom above "bacchanalian revels," the swamp offers them "a fragrant pillow" when they grow tired, beneath a "checkered roof of vine-leaves" (407). The scene recalls the night Jane Eyre passes on the moor with "a low, mossy swell [for] my pillow." In both cases, Nature functions in a time of need as the "mother [who] would lodge me without money and without price."[33]

When Dred discovers the group and brings them to his settlement within the swamp, Stowe radically extends her domestic idea by showing how the swamp serves as a home for a cross-racial, cultivated community: "There are here and there elevated spots in the swampy land, which, by judicious culture, are capable of great productiveness. And many such spots Dred had brought under cultivation" (212). This ability to successfully cultivate spaces within the swamp allows for a removal from a market economy dependent on slavery even as Dred's cultivation efforts render him a successful participant in Lockean liberal individualism.[34] The existence of Maroon communities within North Carolina's Dismal Swamp has been well documented, and according to Herbert Aptheker, evidence supports the existence of at least fifty Maroon communities in the South between 1672 and 1864.[35] Asking the reader to "follow us again to the fastness in the Dismal Swamp," Stowe describes how Tiff and the children are welcomed into the community: the group constructs cabins for the new occupants, Tiff has a sweet potato patch to garden, and the children are found "roaming up and down, looking for autumn flowers and grapes" (446). The settlement is a space in which the land and labor are shared, and Stowe emphasizes the productive and self-sufficient nature of the community. Dred is a kind of Jeffersonian small farmer, revolutionary and self-sustaining, rendering him a model citizen within an interracial polity.

Moreover, the swamp settlement becomes an egalitarian alternative community sheltering a roster of characters that throughout the novel swells to include Harry and Lisette, Tiff and the Cripps children, and Clayton after Tom Gordon's attack. Stowe alludes to Scripture to describe the restorative nature of the environment in aiding Clayton's recovery after an assault:

> As air and heat and water all have benevolent tendency to enter in and fill up a vacuum, so we might fancy the failing vitality of the human system to receive accessions of vigor by being placed in the vicinity of the healthful growths of nature. All the trees which John saw around the river of life and heaven bore healing leaves; and there may be a sense in which the trees of our world bear leaves that are healing both to body and soul. He who hath gone out of the city, sick, disgusted, and wearied, and lain himself down in the forest, under the fatherly shadow of an oak, may have heard this whispered to him in the leafy rustling of a thousand tongues. (508)

Contextualized in this manner, the Maroon community is part and parcel of a nurturing natural world. Distinctions between human and environment dissolve in this passage, as "healthful growths of nature" provide "accessions of vigor." The division between human and landscape becomes porous, and to emphasize this, Stowe points out the "fatherly" protection afforded by an oak, and characterizes the rustling of the leaves as the "whispered" speech of "a thousand tongues." Anthropomorphism here signals nature's capacity to heal and invigorate.

By depicting the Maroon community as a productive and healthful alternative to the plantation, Stowe enters into dialogue with contemporary land use debates. Enslaved people were frequently allowed to keep small gardens, and these spaces were, in practice, their own, though the existence of these patches raised questions about use and ownership. The slave narrator Charles Ball describes in *Slavery in the United States* (1836) how "the people are allowed to make patches, as they are called—that is, gardens, in some remote and unprofitable part of the estate, generally in the woods" for sustenance and profit.[36] Ball notes that this practice was common, highlighting the small-scale economy fueled by slaves laboring in their own gardens. Slave owners could take a surprisingly pragmatic approach in this regard. Thomas Ruffin, the North Carolina Supreme Court judge who infamously ruled in 1829 that owners' authority over the enslaved was total, also weighed in on the rights of slaves to these garden plots. Ruffin, a future president of the North Carolina Agricultural Society, ruled in 1845 that slaves had the right to the small profits from personal gardens, citing numerous reasons for this: the allowance promoted good will among slaves, it saved money for the plantation, and the "little crops" did not constitute property in an economically significant way.[37] This same man had earlier ruled that in cases of slaveholder violence slaves had no right to legal recourse since "the power of the master must be absolute."[38]

Plantation owners frequently came under attack from abolitionists and nonslaveholding southerners who pointed out that the large-scale practice of single-crop agriculture in the South was bad for the economy and, by extension, for their politics. In his 1857 *The Impending Crisis of the South*, for instance, Hinton Helper used charts and statistics to demonstrate how the South lagged behind northern agriculture. Describing the natural "quality and variety" of his own soil, Helper pointed to slavery as the reason why southern states as a whole were not more successful at agricultural enterprises on the national market.[39] After journeying through the coastal slave states, Frederick Law Olmsted came to a similar assessment. In *A Journey in the Seaboard Slave States* (1856) Olmsted describes the "exhausted fields" in Virginia resulting from a "laisser fair [*sic*] principle [that] reigned in their politics," allowing for soil depletion through the overplanting of tobacco.[40] It was this journey that interested Stowe when she wrote to Olmsted asking for details of the vegetation that he had encountered.

Northern farming papers wrote on this topic frequently, naming slavery as the obstacle to southern success at agriculture. William Kendrick's widely republished 1839 article on "Alleged Effects of Slavery on the Agriculture of Virginia," for instance, describes the fertility of the Virginian soil, whose "calcareous manures, for the renovations of these lands, are inexhaustible, and . . . profusely scattered over the whole country, far and wide." But he notes how slavery "degrade[s] . . . the profession" and exclaims that "the people of Virginia will never be able to compete with their brethren of the less highly favored land of New England, either in agriculture, or manufactures, or commerce, until some great change, under providence, can be brought about in the political condition of their people."[41] The system, not the soil, is to blame. In fact, critiques of southern monoculture and slavery entangled agricultural and political discourse.

In *Dred*, Nina proclaims herself an abolitionist right after she expresses her belief that northerners have a higher quality of life despite "stony hills, and poor soil" (151), and Clayton later incites vitriol from the planter class when he gives a speech at the Washington Agricultural Society critiquing current slave laws. A local paper condemns his speech as "Covert Abolitionism!" (467). And in contrast to the general sense of decay that plagues the Gordon estate, Dred's successful cultivation in the swamp emphasizes the richness of southern soil when worked through an alternative agricultural model. The pluralism of the Dred's community—both racial and botanical—is the key to its success.

Horticulture, Hybridity, and the Language of Race

In addition to celebrating both racial and botanical diversity in its vision of a future polity, *Dred* explicitly engages with botany's role in racial formation. Just as botany's sexual classification system gave early Americans a language for talking about sexuality before the rise of sexology in the late nineteenth century, botany and horticulture were deeply enmeshed in understandings of race.[42] This vocabulary shored up the idea of biological difference and hierarchy but stood apart from the infamous work of racial ethnographers who trafficked in the language of animality.[43]

Stowe was writing at a moment when the science of plants was undergoing a tremendous shift. The early American republic had an insatiable desire for new plants and actively traded with Europe for novel varietals. Yet by the early decades of the nineteenth century, the practice of plant breeding, fueled by growing knowledge about the sexual nature of plant reproduction, offered another means of creating variety through crossing. The popularization of botanical experimentation carried with it the important idea that humans could generate, rather than merely observe or use, organic matter.

Perhaps one of the most famous examples of horticulture indexing race occurs in James McCune Smith's 1855 preface to Frederick Douglass's *My Bondage and My Freedom*, when McCune Smith invokes the language of the graft to frame Douglass's prodigious gifts. As we saw in chapter 1, the graft provided Lydia Maria Child and others with a potent symbol for the Americanization of Europeans. But McCune Smith makes grafting an explicitly racial metaphor. He writes of Douglass that "the versatility of talent which he wields, in common with Dumas, Ira Aldridge, and Miss Greenfield, would seem to be the result of the grafting of the Anglo-Saxon on good, original, negro stock."[44] In a national culture obsessed with origins, McCune Smith also subtly reminds the reader that "Anglo-Saxon" is a transplant on American soil, one grafted rather than rooted.

Like grafting, plant hybridizing did not require a specialist and was readily accessible to anyone familiar with the basic principles of gardening. Domestic and horticultural publications noted that the cross-fertilization of flowers was a pastime befitting the lady gardener at home.[45] Writers foregrounded the possibility of controlled selection for desired traits by cross-pollinating varieties from different countries. "The flowers of our common Honeysuckle are little more than an inch long," the English botanist John Lindley wrote in 1847, "but those of the beautiful Tuscan species (*Caprifolium etruscum*) are twice as large, and in the Caprifolium longiflorum of China the blossoms are full three inches in length. Here are the best opportunities for an improvement of one of our most favorite plants."[46] Lindley's suggestion meant that individual gardeners at home could radically intervene in nature for a desired result. Seedlings from around the globe were readily available by midcentury as the circulation of plants developed into a commercial industry, and home gardeners were invited to experiment with the results. Lindley closes with this reassurance: "The plants that are adverted to are all of common occurrence, and easy for any one to operate upon . . . and they offer an abundant reward to skill and patience."[47] At home, anyone could create a new cross based on seedlings from around the globe.

The process of creating hybrids encouraged a profound shift in thinking about the natural world and the role of humans in intervening in the natural order. Unsurprisingly, the greatest opponents to hybridization tended to be religious figures who felt that such an intervention disturbed God's natural plan. While Stowe retained a view of God's order at work in the natural world, her language provides a more accommodating relationship between her theological convictions and science. In a passage in *Dred* that sounds presciently like a discussion of Gregor Mendel's peas, Stowe engages the question of nature versus nurture with respect to the traits expressed by her characters. Describing Clayton's relationship to his father, she writes, "Between Clayton and his father there existed an affection deep and entire; yet, as the son developed to manhood, it became increasingly evident that they could never move harmoniously in the same practical orbit. The nature of the son was so veined and crossed with that of his mother, that the father, in attempting the age-long and often-tried experiment of making his child an exact copy of himself, found himself extremely puzzled and confused in the operation" (27). Stowe mocks Clayton's father's desire for an exact replica, and in describing this lineage, Stowe emphasizes that human biology is combinatory—an idea widely accepted in the realm of plant culture well before the discovery of genes. The connection between botanical and human reproduction would have been evident to any nineteenth-century reader familiar with Linnaeus's claim that "the reproductive parts of plants paralleled the sex organs of animals."[48] Indeed, as Elise Lemire notes, the racial application of the word "cross" originates in the language of horticulture.[49]

The creation of plant hybrids, like the desire to discover new species in the wilderness, grew from a conscious, progressive effort, clearly articulated in an 1848 volume of the *Eclectic Magazine of Foreign Literature*, a long-running British magazine that frequently reviewed American literature. In an article titled "Pleasure of Botany and Gardening," one horticultural author writes that "it was not suspected till lately, that while vegetable life can only be called into existence by the Divine Artificer, it is allowed to his servant, man, to turn that life into new channels, and to impress upon it forms of beauty unknown and unseen before. . . . It is by hybridizing that art achieves its most exalted triumphs in this department of nature."[50] Here experimentation is rendered as serving, rather than meddling with, a divine plan. The idea of biotic improvement held cultural sway at midcentury and, in *Dred*, became a way for Stowe to express her progressing views on race and her revisioning of a democratic society.

This point is most readily apparent in the racist impulses of the Clayton siblings. At a key moment near the end of the novel, Edward Clayton and his sister Anne discuss the possibilities for the future of a postslavery society, and their conversation about Black education is full of rigid racial classifications. The siblings first agree that "the African race evidently are made to excel in that department which lies between the sensuous and the intellectual—what we call the elegant arts" (328). Elaborating, Clayton uses a botanical metaphor for racial progress. He tells Anne, "The Ethiopian race is a slow-growing plant, like the aloe . . . but I hope some of these days, they'll come into flower, and I think, if they ever do, the blossoming will be gorgeous" (328). He then goes on to insist upon classificatory distinction between Blacks and whites: "There is no use trying to make the negroes into Anglo-Saxons, any more than making a grape-vine into a pear tree. I train the grape-vine" (328). Samuel Otter rightly notes the "peculiar" quality of this moment of "horticultural condescension," pointing out that this passage is "contradicted, or at least distanced, by the narrator's quite different account of Dred's extravagant perspective."[51] Framed in relation to the rest of the novel's botanical depictions, Clayton's analogy echoes the reductive use of botanical language by one of Nina's failed suitors. In an important earlier scene that teaches us to be wary of reading flora in classificatory terms, a man named Carson attempts to woo Nina by giving her a "full-blown rose" and asking if she understands its meaning within the codified language of flowers. Nina, in response, "pluck[s] from the bush a rose of two or three days' bloom, whose leaves were falling out, . . . hand[s] it to him," and says, "'Do you understand the signification of this?'" (120). Stowe is here quick to parody the transparent social codes that sentimental flower books purported to explicate, and she points out that botanical science could be just as reductive. Like Carson, Clayton thinks analogically as he seeks to ground his essentialist views in a rigid classificatory system, but the novel engages such systems only to destabilize their authenticity.

Importantly, Clayton's racist metaphor conjures a debate about plant transmutation that occurred through the first half of the century—the theory that one plant could turn into another.[52] The subtext of this debate is the nature of succession in the natural world, a theme that Thoreau would later speak to passionately in "The Succession of Forest Trees" (1860). Left to their own devices, horticulturalists observed, plants naturally replace one another so that "different tribes of vegetables do *succeed* and *supersede* each other, in our fields and meadows."[53] This observation provides the subtext to Clayton's commentary, suggesting that, as Ian Frederick Finseth notes, the natural world was a central concept in the battle over slavery in the 1850s because it serves as "the very foundation on which ideas of race were grounded."[54] Clayton's botanical metaphors render race as a biological fact, but his analogy also invites readers to consider the possibility of racial succession.

At the very least, Clayton's choice of plants from the garden undercuts the hierarchy he establishes. The two fruits he selects, the pear and the grape, were politically important crops in the late eighteenth and nineteenth centuries because the cultivation of each participated in an international narrative about the productivity of American soil. Philip Pauly has described how "American independence was—and was understood at the time to be—a biohistorical event," and thinkers considered "the degree to which the United States was a unique climatic region, and the extent to which cultured Old World plants could be naturalized in the United States" as signs of the republic's potential for success. Grapevines and pear trees were both imported because of their association with Europe, yet each posed problems for horticulturalists in America. By 1800, as Pauly notes, the import of European grapevines had been curtailed because of a high failure rate, but the search for a successful native grape led to the cultivation of several indigenous varieties. The Catawba had begun to thrive in the 1840s in southwestern Ohio, years when Stowe was living in Cincinnati, and the Concord was discovered in Massachusetts during the same decade. The growing success of grape culture by midcentury contrasted with a decline in pear culture over the course of the same decades. The pear, which, as Pauly notes, "Downing and Thoreau satirized . . . as foreign and pompous," suffered from blight so severe that the whole enterprise was called into question.[55] So it is notable that, at a time when racist discourse argued that Black people were dying out in America because they were unsuited to the climate, Clayton compares Black people with the more successful plant culture—the grape, and not the pear. In doing so, Clayton undercuts the botanical hierarchy he attempts to establish when asserting the superiority of white culture. Indeed, Stowe demonstrates the vitality of the native grapevine in her description of the swamp. Noting their appearance surrounding the Maroon community, she describes how "the wild climbing grapevines, which hang in thousand-fold festoons round the enclosure, are purpling with grapes" (445). Not only does Clayton's image create an ironic hierarchy, but the choice of grapevines and pear trees naturalizes Blacks on American soil while associating the Anglo-Saxon with a dying foreign culture.

Elsewhere Stowe uses geography quite differently. Accounting for Dred's close identification with biblical stories, the narrator reflects, "Even in the soil of the cool Saxon heart the Bible has thrown out its roots with an all-pervading energy, so that the whole frame-work of society may be said to rest on soil held together by its fibres. . . . But, when this oriental seed, an exotic among us, is planted back in the fiery soil of a tropical heart, it bursts forth with an incalculable ardor of growth" (211). Stowe's shifting horticultural rhetoric indicates an ambivalent, if not quite ironic, attitude toward essentialism. Her figuration of the Bible renders Christianity an "oriental seed, an exotic among us." By suggesting that the "fiery soil" of Dred's "tropical heart" will allow it to thrive, Stowe makes Dred's presence on American soil a key to the nation's success.

As the failed plant experiments in Jefferson's garden demonstrated, the success or failure of gardening ventures served as a perceived litmus test for the potential of the new republic. The great lengths to which gardeners went to maintain tropical plants in the cold, damp climates of Britain and New England indicate the cultural investment in such practices. For example, the successful cultivation of the enormous South

American *Victoria Regia* water lily in England in 1850 led the American *Horticulturalist* to express hope that the president of the Pennsylvania Horticultural Society would soon construct an appropriate greenhouse to follow suit.[56] Whereas Jefferson's *Notes on the State of Virginia* worried about the relationship between horticultural and democratic experiments, Stowe saw in horticultural diversity an answer to the flaws of the current political model that monopolized rights for one segment of the population. Dred's association with "an incalculable ardor of [religious] growth" rewrites the script for the revolutionary gardener. His religious empathy with the plants around him renders him a successful cultivator, and he models an existence that is in communion with, rather than control of, the landscape around him. Stowe favored the resistance to hierarchy implied in a sympathetic relationship to plants and believed that the practice of gardening could influence the gardener in this regard.

* * *

Stowe's botanical rhetoric asserts that the naturalist's perspective on the world provides moral ballast and collapses binary thinking while upholding a principle of growth. The wilderness in *Dred* is always, in some sense, already incorporated into the domestic imaginary where its anarchic beauty is celebrated.[57] This understanding of the domestic space emphasizes vitality in excess of management. In "The Succession of Forest Trees," Thoreau delights in the naturalist's freedom to explore beyond property boundaries. His justification for speaking to a group of farmers lies in his intimate knowledge of their lands; as he writes, "taking a surveyor's and a naturalist's liberty, I have been in the habit of going across your lots much oftener than is usual, as many of you, perhaps to your sorrow, are aware," adding, "I have several times shown the proprietor the shortest way out of his wood-lot."[58] To be a naturalist in this vein is to collapse the boundaries of ownership and seek a different kind of ordering in the world through a sensibility oriented by organic interconnection.

The growing sectional divides on the eve of the Civil War vexed the issue of national growth, as geographical expansion was deeply entangled with debates about slavery. At the same time, shifting botanical practices provided new meaning to the rhetoric of personal growth through cultivation. Stowe draws on this potent cultural trope because it relates at once to the universal and the very local. Plants provided communion across boundaries—of region, race, class, and gender—while also functioning as fundamental features of the domestic space. Such recourse to mobility and stability claims the home as a porous space, and Stowe invites the reader to think of the world not in terms of neatly drawn boundaries, but rather, expansively, discursively, and multiply.

Stowe's second antislavery novel does not provide a utopian vision of a postslavery society in expressly political terms. At the end of *Dred*, all of the novel's surviving Black characters have fled to the Northeast or Canada and the institution of slavery continues without abatement. Moreover, even those who gain freedom are conscripted into self-sacrificing roles. As Elizabeth Duquette has noted, at the end of the novel Stowe promotes sacrifice as the condition for Black women's participation in civic society. Yet while Stowe no doubt had a constrained vision of what freedom was and how it could be achieved, her perspective is valuable insofar as her writing illustrates the salience of horticulture, not simply agriculture, to debates about the nation's future. As Stowe's novel makes clear, the discourses of domestic cultivation and agricultural cultivation overlapped in potent ways. In Canada, the "improvement" of Clayton's emancipated slaves tracks with the "improvement" of the soil. But agricultural and domestic cultivation projects also diverged. By making botanical rhetoric mutable, Stowe points to the kinds of growth that defy scientific and sentimental epistemologies.

Attention to Stowe's botanical language reveals how environmental sensibilities affected racial politics in the antebellum decades. Botany helped popularize the act of looking closely at the physical world and held out the promise of greater moral purchase in doing so. Yet all ways of looking are not the same, and Stowe's approach in *Dred* looks and feels different from the common classificatory practices that subtended arguments of natural hierarchy. Stowe's sentimental science rejected the essentially fixed environmental order common to natural theology. In presenting the radical potential of plant growth to cut across manmade boundaries, Stowe described a world that is not fixed, but fluid. In foregrounding botanical variety, she celebrated diversity. The social applications of Stowe's botanical ethic were themselves multiple: if the order of the natural world is not predetermined, then essentialist constructions of race cannot find traction, botanical science emerges as an engine of change, and sympathetic communion with nature might guide growth without exploitation.

Several weeks after Stowe's "Meditations from Our Garden Seat" ran in the *Independent*, Frederick Douglass reprinted it in his eponymous newspaper. Like Stowe, Douglass recognized how the sentimental dimensions of plant culture might serve abolitionist ends. Yet if Stowe had a dismal prognostication for the fate of southern plantations, depicting them as exhausted and worn down, Douglass paid closer attention to the planters' own hopes that new forms of agricultural science could rejuvenate exhausted soil. As we will see in the next chapter, agricultural "improvement" had vastly different implications in slaveholding territory than in the free North.

5

Botanical Societies

In October 1849 Frederick Douglass published a short article in the *North Star* on pumpkins. If the vegetal subject matter seems far afield from a prominent abolitionist, political activist, and orator, Douglass is quick to encourage the skeptical reader. "Yes! Pumpkins!" the article begins. "We raised a nice lot of them this season in our own garden."[1] After characterizing his remarkable homegrown pumpkins for a few lines, Douglass makes an explicit moral turn.

> It is not so much the good *quality* of the pumpkins to which we would call attention, as to the good *moral* we have extracted from them.—The ground was prepared—seed sown—and the plant cultivated by our own colored hands; and although the soil is *American*, it took no offense on account of our color—but yielded a generous return for our industry. From this we infer that the earth has no prejudice against color, and that nature is no respecter of persons. It pours its treasures as liberally into the lap of colored industry, as into that of the white husbandman. The earth is a preacher of righteousness; it inculcates justice—love—and mercy; repudiates the factitious distinctions of pride and prejudice—and owns all the sons and daughters of men (without regard to color) as its own dear children.

Douglass here finds recourse not in the promise of equality set by paper or parchment, as in his famous address "What to the Slave Is the Fourth of July?" (1852), but in the earth, which he argues has a moral logic greater than nation. "Although the soil is *American*," it does not abide by American law, bestowing produce equally to whoever will cultivate it.

The egalitarian vision of nature that Douglass lays out here sharply juxtaposes with the agricultural system that relied upon the relentless toil of enslaved laborers. It stands in direct contrast to the ways in which African American industry in the antebellum period most often yielded a generous return for white slave holders. And it also anticipates and refutes the racist imagery of Thomas Carlyle's "Occasional Discourse on the Negro Question" (1849), which would appear later the same year. Carlyle's incendiary essay rehearses practically every gross claim to Black racial inferiority, mounting an argument based on scientific racism as well as economics. In the essay Carlyle's thinly veiled narrator uses pumpkins to characterize Black cultivation in the West Indies after emancipation, contrasting it with rotting sugar crops as he complains about the shortage of enslaved labor:

> Where a black man, by working half an hour a day (such is the calculation), can supply himself, by aid of sun and soil with as much pumpkin as will suffice, he is likely to be a little stiff to raise into hard work! Supply and demand, which, science says, should be brought to bear on him, have an up-hill task of it with such a man. Strong sun supplied itself gratis— rich soil, in those unpeopled or half-peopled regions, almost gratis: these are *his* supply; and half an hour a day, directed upon these, will produce pumpkin, which is his "demand." The fortunate black man! very swiftly does he settle his account with supply and demand; not so swiftly the less fortunate white man of these tropical localities. He, himself, cannot work; and his black neighbor, rich in pumpkins, is in no haste to help him. Sunk to the ears in pumpkin, imbibing saccharine juices, and much at his ease in the creation, he can listen to the less fortunate white man's "demand," and take his own time in supplying it. Higher wages, massa; higher, for your cane crop cannot wait; still higher—till no conceivable opulence of cane crop will cover such wages! In Demerara, as I read in the blue book of last year, the cane crop, far and wide, stands rotting; the fortunate black gentleman; strong in their pumpkins, have all struck till the "demand" rise a little.

Carlyle returns to the imagery of the pumpkin obsessively across the essay, evoking the fruit thirty-eight times to figure liberated Black cultivation and to underscore his disdain for the way emancipation restructured economic and social relations. At one point he stresses that raising pumpkins is a waste of productive soil that would better be used in producing refined commodities like "sugars, cinnamons, and nobler products." Later he argues that white men were responsible for turning "the jungles, the putrescences and waste savageries" into arable land, and so therefore any products grown on the islands are rightfully the fruit of white "heroism."[2]

Douglass's article offers a window into the ways in which cultivation was often at the nexus of slavery debates that were at once economic (Carlyle's coinage of economics as the "dismal science" comes from this essay) and moral. The noncommodified pumpkin, which becomes a symbol for Carlyle's fury, is in Douglass's hands a symbol of pride. The moral lesson yielded from "our own" garden hinges on this possessive phrase; plenty of land was "prepared—seeds sown—and the plant cultivated by our own colored hands" that yielded "a generous return for our industry" that went directly into the pockets of white owners. Douglass's articulation of "our own garden"—one whose yield benefits the laboring gardener—by contrast emphasizes both the self-possession of the laborer and claim to the fruits, or vegetables, of one's own labor. In emphasizing that the earth is indifferent to color, Douglass derives an ecological response to proslavery arguments that were based in supposedly "natural" difference.[3] But recourse to "our own garden" is also part of Douglass's effort to address the nexus of cultivation, plant life, and property as it relates to the potential for Black prosperity. His engagement with horticulture anticipates a long genealogy of Black writers and intellectuals addressing the meaning of plant cultivation in the wake of slavery.[4]

Douglass's approach to cultivation is particularly significant for the way that it establishes the relevance of scientific agriculture to the environmental politics of slavery and freedom. The importance of environmental thought to Douglass's fight for justice has gained increasing attention in recent years. As political scientist Kimberly Smith has pointed out, Douglass is part of a rich vein of African American environmental writing in the nineteenth century that for many years went unrecognized.[5] Michael Bennett has described the antipastoralism of Douglass's earliest autobiography as his means of critiquing the how the pastoral had been used in the perpetuation of white supremacy.[6] And Cristin Ellis has shown how Douglass, among others, leveraged the environmental crisis of southern soil exhaustion to make a pragmatic bid for abolition. My reading builds on these accounts to locate how scientific agriculture shaped environmental debates about slavery. In particular, the scientific study of farming practices from the 1840s until the 1870s changed methods of farming at all scales. Across the nineteenth century the discourse around farming became increasingly empirical to the point where educational reformer Josiah Holbrook, in an 1853 article for the *National Era* on the "Democracy of Science," called agriculture "the science of all sciences."[7]

As farming practices became more scientific, writers increasingly explored the implication of this science on local environmental conditions and the South's political economy. In *Cultivation and Catastrophe* (2017) Sonya Posmentier elucidates the significance of the split in the ecological perspectives of W. E. B. Du Bois and Booker T. Washington, identifying how their different understandings of cultivation—as alienation versus intimacy—"mark[ed] the limits of the range of possibilities that become available to later writers attempting to inhabit, recover, or transform the geography of the plantation."[8] In the antebellum period, writers grappling with the significance of Black cultivation did so against the immediate backdrop of plantation slavery and a growing sectional split in agricultural organization. Douglass, who had performed agricultural labor in addition to other forms of work while enslaved, spent years in Rochester, New York, where agricultural interests became increasingly organized. His sense of cultivation is shaped by an understanding of its role in plantation agriculture, the alternative horticultural practices of his grandmother and other enslaved people, and the discourse about scientific improvement among a rapidly developing northern agricultural network. Douglass's depiction of various methods of cultivation acknowledges both its alienating effects and its potential as a tool for Black economic and social advancement.

If agrarianism is an ethos with deep roots in the American environmental psyche, what it means to farm takes on a new context related to self-sufficiency and community for free Blacks in the antebellum period, during the war, and especially after. Douglass joined Stowe in exploring horticulture as a form of civic engagement in antebellum America, but he was particularly attentive to how developments in scientific agriculture shaped both what Paul Sutter calls plantation "agroecosystems" and alternative methods of working the land.[9] In his writings Douglass stressed what many implicitly recognized: horticulture could be a civic tool, but it could also help maintain the civil veneer of the brutal regime.

Douglass's career as a public orator, newspaper editor, and prominent activist gave him a unique platform for showing how developments in horticulture could be useful to slaveholder and abolitionist alike.

Douglass and Southern Agrarianism

By the early decades of the nineteenth century soil exhaustion was one of the most pressing threats to southern agricultural profit. Decades of unsustainable practices, including the widespread practice of monoculture, had led to depleted land and profits. Abolitionists were quick to capitalize upon this crisis and made soil infertility the basis for moral, economic, and political antislavery arguments. The nature of the crisis was the subject of conversation in both the North and South, and even led some in the South to renounce slavery.[10] Yet if soil exhaustion was an active worry among the southern planter class by the mid-nineteenth century, it was not met with resignation. Many planters seeking to preserve their way of life put stock in the newly developing field of chemical land management. Southerners and northerners alike heralded guano—seabird and bat excrement high in nitrogen and phosphorus—as a remedy for exhausted soil.[11] The application of guano was only one small part of larger developments in agricultural practice toward chemical appraisal of the soil and faith in scientific management. Planters, whose deeply cherished way of life was threatened by crop failure, sought solutions in agricultural innovations.[12] Many participated in agricultural societies designed to improve yield and stave off infertility in order to maintain the plantation system.[13] And as scientific management increased, so too did actuarial developments in the way planters calculated profits. As Caitlin Rosenthal puts it, "Innovation and brutality went hand in hand."[14]

The Lloyd plantation as Douglass depicts it across his biographies was part of a network of scientific improvement. One way to understand the Lloyd plantation is through its investment in "pastoral seclusion" maintained by "feat of protectionist engineering."[15] Douglass, as Ellis has convincingly shown, exposes the violence that belies this pastoral idyll. But for all their ability to engineer seclusion, planters were not isolationist insofar as they were deeply invested in horticultural innovations and methods that would serve their interests. This is apparent in both *Narrative* and *My Bondage* in the way Douglass characterizes the property's "large and finely cultivated" garden.[16]

Critics often read Colonel Lloyd's garden in his early autobiography *Narrative* in biblical terms, for good reason.[17] As Douglass recounts, "The colonel had to resort to all kinds of stratagems to keep his slaves out of the garden. The last and most successful one was that of tarring his fence all around; after which, if a slave was caught with any tar upon his person, it was deemed sufficient proof that he had either been into the garden, or had tried to get in. In either case, he was severely whipped by the chief gardener" (29). As the garden was a restricted space for any enslaved person who wasn't explicitly assigned to work there, the tarring recalls the story of Cain and Abel. In contrast, the garden registers as a kind of prelapsarian Eden for white visitors. Douglass characterizes it as "probably the greatest attraction of the place," noting its role in attracting visitors to the plantation: "During the summer months, people came from far and near—from Baltimore, Easton, and Annapolis—to see it" (29). The garden here becomes a public display, legitimating the activities of the planter through what Richard Drayton has called elsewhere "the aestheticization of power."[18]

In *My Bondage and My Freedom* Douglass goes further to express the garden as a site of cutting-edge scientific agricultural knowledge. Rather than suggesting that the Lloyds adopted a parochial stance toward soil health, Douglass emphasizes the way in which the plantation's bounty was supported by scientific methods. The passage on the garden in *My Bondage* begins by contrasting the opulence of the Lloyd table with the inadequacy of the diet of the enslaved.[19] But his revision also situates Lloyd's garden—and its gardeners—within a scientific context. In *My Bondage* the policing of the plantation garden ensures that the Lloyd table "groans under the heavy and blood-bought luxuries gathered with pains-taking care."[20] After noting "the close-fisted stinginess that fed the poor slave on coarse corn-meal and tainted meat" he describes "a fertile garden, many acres in size, constituting a separate establishment, distinct from the common farm—with its scientific gardener, imported from Scotland, (a Mr. McDermott,) with four men under his direction."[21] That McDermott has been "imported" from Scotland in Douglass's characterization wryly highlights the degree to which he is a fungible representative of his type—the scientific gardener—whose services a wealthy planter class family could afford to purchase. Like Scotland's textiles and machinery, and indeed somewhat like the enslaved men "under his direction," McDermott circulates with market demand. But Douglass's addition of this detail also matters because it suggests that the Lloyds were not isolationist in their cultivation methods but rather were part of a larger network actively pursuing scientific means of managing the garden.

The family's library, for instance, reflected an interest in scientific gardening dating back to the eighteenth century, and the garden showed investment in scientific instruments and technologies available to wealthy landowners. The greenhouse, installed in the late eighteenth century, included a furnace, water irrigation system, and thermometer.[22] These technologies made it easier to grow plants from a range of different climes in defiance of local climatological difficulties.

The fruits of scientific gardening on the Wye plantation, Douglass notes, were copious and attractive. He lists the produce, emphasizing how "tender," "succulent" and "delicate" is the yield of the garden, in contrast to the provisions allowed the enslaved. The garden's produce, he enumerates, includes "tender asparagus, the succulent celery, and the delicate cauliflower; egg plants, beets, lettuce, parsnips, peas, and French beans, early and late; radishes, cantaloupes, melons of all kinds; the fruits and flowers of climes and all descriptions, from the hardy apple of the north, to the lemon and orange of the south, culminated at this point."[23] The goal of scientific gardening was, broadly speaking, to increase yield, experiment with new varieties, and create such a nexus where fruits from southern and northern climes might be grown equally well. The prodigious culmination of northern and southern fruits that Douglass describes stands in sharp contrast to the kind of monoculture familiarly ascribed to southern plantations, and the kind that might exist on the plantation's "common farm" that he mentions in passing. Douglass goes to pains here to underscore that the garden is at a remove from this part of the property and its methods, emphasizing that the garden is so wholly distinct as to constitute a "separate establishment." This allows for the Lloyds to assume psychological as well as physical distance from agricultural labor, and in distinguishing the garden from the farm, Douglass nonetheless emphasizes its role in sustaining the plantation regime. For all that the garden is separated from the monoculture nearby, it illustrates horticultural expertise in service of one family's groaning table.[24]

Ellis has productively suggested that we take Douglass's descriptions of slaves as a "human crop" as more than a rhetorical dig at planters' dehumanizing impulse, pointing out that by the 1840s soil exhaustion meant that "slaves had become a more valuable product to the South than the cash crops to which it still staked its regional pride" (276). The depletion of the soil and the crisis it augured were increasingly hard for planters to deny, and Douglass recognized the power of a systemic tact to dismantling the economic logic of slavery.[25] At the same time, Douglass also expressed anxiety about the ways that planters were adapting to fertility issues. For if soil exhaustion was a powerful argument against monoculture, this argument was potentially rendered moot by new technologies and new applications to rejuvenate the soil. Chemical fertilizers, novel varieties of crops, and other innovations of scientific agriculture could potentially offer planters the solutions to grow crops using monoculture and enslaved labor in perpetuity.

Articles in the Richmond-based journal the *Southern Planter* from the 1850s illustrate the prevalence of discussion around scientific agriculture and soil amendments among its audience of planters and small farmers in the Upper South. While soil improvement had been a topic of experimentation and debate since the revolutionary period, the *Planter*, which began its run in 1841, was part of an increasingly robust effort by midcentury to organize a network of agriculturalists along these principles.[26] That is, despite the crisis in soil fertility, many planters remained optimistic by putting their faith in innovation. One reader's editorial, titled "Superiority of the Farming Profession," lauds the happiness and success of planters as a class, urging others to take up the mantle: "Look around through the land, and see who are the most independently settled young men in the commonwealth, and I venture to say, that with an occasional exception, they will be found to be the sons of planters." Yet even in advocating this path, the author, referred to only as "South Side," makes clear the requirements: "a soft, fertile, generous soil, with all the necessary appliances to make it productive, and a few tons of guano to quicken its failing energies."[27] Several pages later, another article in the same issue focuses on "Chemistry Applied to the Mechanic and Farmer." The author begins by stressing that chemical laws govern everything, "as unchangeable as those which hold the planets in their places." After illustrating how the sandstone exterior of the U.S. Capitol building is "being rapidly acted upon and crumbled to dust by the common atmospheric agents" (206), the author is keen to show how similar principles apply to the farm field. "The farm is a great laboratory, and all these changes in matter which it is the farmer's chief business to produce, are of chemical nature." Whereas preserving the crumbling Capitol building presents a complex challenge, the field can be more easily remediated through chemical application. This focus on chemistry in some respects makes the scale of farming irrelevant: measure input correctly and output should be predictable. If abolitionists were quickly coming around to attacking plantation monoculture on the basis of soil exhaustion, the *Planter* sought other explanations. Flagging soils mean a chemical problem, not a labor one. The article chides "those farmers who decline to inquire into the principles which govern their vocation, or who prefer the study of politics to that of agriculture," explaining that they "will have occasion to groan more deeply than ever over the unprofitableness of their business."[28] Of course, as sectional tension increased throughout the 1850s, articles and editorials in the *Planter* became increasingly explicit about their proslavery politics.

Over a decade after Emancipation, Douglass again nodded to the imbrication of agriculture and politics in the language he used to characterize the District of Columbia. Speaking in his new role as federal marshal for the district, Douglass outlines the relationship between slavery and agricultural technology in "A Lecture on Our National Capital." After praising the city for its layout, infrastructure, private estates, and public greenways, Douglass critiques its siting using an agricultural metaphor: "Looking at the influence exerted by simple local surroundings, I have no hesitation in saying that the selection of Washington as the National Capital was one of the greatest mistakes made by the fathers of the Republic. The seat of government ought never to have been planted there." Referencing the influence of the planter class, he continues, "Sandwiched between two of the oldest slave states, each of which was a nursery and hot-bed of slavery; surrounded by a people accustomed to look upon the youthful members of a colored man's family as a part of the annual crop for the market; pervaded by the manners, morals, politics, and religion peculiar to a slaveholding community, the inhabitants of the National

Capital were, from first to last, frantically and fanatically sectional."[29] What is striking here from a horticultural perspective is not simply that Douglass reiterates the idea of the human crop that Ellis notes he uses in *Narrative*. The picture that Douglass elaborates is also explicitly connected to horticultural technologies that allowed gardeners to control environmental conditions. The nursery and the hotbed are both means to promote growth by buffering plants from unpredictable weather. As metaphors they nod not only to the history of slavery and the political protection it received but also to the ways in which technological improvement maximized human and plant growth for profit.

Horticulture and the Slave Narrative

Left out of the southern agricultural journals, unsurprisingly, were any references to the horticultural expertise or experience of the enslaved. As Douglass's reference to the Lloyd garden indicates, however, enslaved workers were responsible for maintaining this crown jewel of the plantation, in addition to the productivity of the fields. Douglass was one of several prominent Black abolitionists who invoked the decorous horticultural estate as a way to highlight the realities of plantation slavery. Such narratives provide insights into the kinds of horticultural knowledge of the enslaved while also signaling the emancipatory limits to the genteel concept of a garden politic that did not address—or often even acknowledge—the impact of white supremacy on horticultural ideas of the nation.

In the opening paragraph of *Narrative*, Douglass recounts the difficulty of discerning his own birthday, noting that few slaves could get closer to determining the date of their birth than "planting-time, harvest-time, cherry-time, spring-time, or fall-time" (17). The absence of any further certitude signals just one of many deprivations of slavery. But it also illustrates the ways in which planting cycles ruled the lives of agricultural workers, whose days and years were organized around annual crop cycles.

As noted in chapter 4, many plantation owners across the South allowed their enslaved laborers to cultivate small patches of land for their own nourishment, relieving the planter of the need to feed their workers sufficiently. Such patches often both sustained slave families by supplementing meagre plantation rations and supported the growth of city markets.[30] In revising his autobiography for the publication of *My Bondage and My Freedom* in 1855, Douglass added details about his grandmother's horticultural expertise. Grandmother Bailey was "more provident than most of her neighbors in the preservation of seedling sweet potatoes" and respected for miles around for her fishing skills and horticultural prowess. What neighbors called her "good luck," Douglass emphasizes, "was owing to the exceeding care which she took in preventing the succulent root from getting bruised in the digging, and in placing it beyond the reach of frost, by actually burying it under the hearth of her cabin during the winter months." For this reason she was consequently "sent for in all directions, simply to place the seedling potatoes in the hills." Yet neither her years of service nor her skills protect her from being treated abysmally as chattel, and as Douglass movingly notes, in her old age she is turned out from her home, expected to live out the remainder of her days at an isolated cabin in the woods.

Like Douglass, abolitionist Solomon Northup highlighted that horticultural expertise was not only the provenance of the white elite. In his 1853 narrative *Twelve Years a Slave* Northup rehearses his debasement from being an enterprising farmer of his own fields, "an occupation congenial to my tastes" (49), to an enslaved worker in Louisiana. When Northup, a free Black northerner, was illegally sold into slavery and sent south, he spent twelve years working in cotton and cane fields, and his subsequent narrative elaborates on the nature of this work in detail. Such descriptions add the "authenticity" that Benjamin Quarles has argued was vital to the slave narrative genre as a tool in the abolitionist arsenal.[31] But Northup's aesthetic descriptions make a further point, oscillating between scenes of beauty and brutality to show how entwined the aesthetics of the plantation were with extreme violence. Describing the vista of a cotton field, he writes, "There are few sights more pleasant to the eye, than a wide cotton field when it is in bloom. It presents an appearance of purity, like an immaculate expanse of light, new-fallen snow."[32] This reflection comes in the midst of a description of the extreme working conditions in the field and stands in sharp tonal contrast to the descriptions of labor that immediately precede and follow it. Two lines earlier Northup describes a fellow enslaved worker who "would surely have been beaten" if she failed to pick the expected measure. And immediately following the tranquil image of the blooming landscape, Northup describes the "severest chastisement" that awaited any enslaved worker who accidentally broke a branch off the cotton stocks (97). Buttressed in this way, Northup's comment on the beauty of the field must be read with a heavy dose of irony. This pastoral interlude engages romantic aesthetics but refuses to detach such a view from the brutality that sustains it.

Northup further underscores his horticultural knowledge in describing the plantation gardens he encounters at the home of his first master in Louisiana. When financial difficulties led his first master, William Ford, to sell him to a "small, crabbed, quick-tempered, spiteful" carpenter named Tibeats, the man's violent nature spurs Northup to return to Ford for protection (63). After a harrowing escape through snake- and alligator-infested swamp water, Northup arrives at Ford's plantation and recounts the abuse he's suffered. The next day, he "strolled into the madam's garden" and describes a scene of late-season beauty, noting that "though it was a season of the year when . . . the trees are stripped of their summer glories in more frigid climes, yet the whole variety of roses were then blooming there, and the long, luxuriant vines creeping over the frames." Moreover, "The crimson and golden fruit hung half hidden amongst the younger and older blossoms of the peach, the orange, the plum, and the pomegranate" (86). Northup here signals both the geographic range of his horticultural purview, registering the fact that trees in the North would be stripped bare at this point in the year, and his appreciation for the garden's beauty. Immediately after this reflection he devotes time to trimming and weeding the garden as a way to "repay [the Fords'] kindness" (86). In lingering on the prodigious scene, Northup signals his own aesthetic appreciation of the local environment, and his discerning eye in comparing it to northern climes illustrates his familiarity and expertise.

Northup's work in the garden also reminds the reader that the Fords continue to afford this lush scene because they sold him. The beauty of the scene is a temporary refuge that stresses the vast gulf between his situation and that of the Fords. In a similar vein, Harriet Jacobs turns to botanical language to underline the vastly different realities experienced by young women across the color line. Floral language, as we've seen in earlier chapters, had long been associated with notions of moral propriety and white gentility, as sentimental ideology held as its ideal the "blooming" and virtuous young white woman. Jacobs parodies this symbolic floral language in *Incidents in the Life of a Slave Girl* (1861). Describing a scene in which two girls are playing together, one white, the other "her slave, and also her sister," Jacobs intimates the difference in path between the "fair child" who "from childhood to womanhood [had a] pathway [that] was blooming with flowers, and overarched by a sunny sky" and her slave sister who "was also very beautiful; but the flowers and sunshine of love were not for her."[33] The floral language stresses the vastly divergent trajectories of the sisters based on their racial designations, underscoring the way in which sentimental floral language operated within a racial matrix. Several pages later Jacobs continues with the floral descriptors to further accentuate the corruption at the heart of the southern pastoral myth. Northern men, she notes, are "proud" to marry their daughters to slaveholders, and "the poor girls have romantic notions of a sunny clime, and of the flowering vines that all the year round shade a happy home." This romance is punctured by husbands who rape enslaved women, letting "jealousy and hatred enter the flowery home" and leaving it "ravaged of its loveliness."[34]

While the southern home and garden presented a powerful site for sentimental critique, abolitionists also looked to Europe for other models of agriculture and agricultural labor in what Martha Schoolman has called a "routine gestur[e] of antislavery transatlantic political comparison."[35] Novelist and intellectual William Wells Brown, who escaped from slavery in 1834, included such a discussion in his travel narrative, *Three Years in Europe*. A number of critics have characterized *Three Years* as developing a mode of "fugitive tourism" that emphasizes "a mobility that is profoundly different from the forced movements of the middle passage and the domestic slave trade, and different too from the 'covert geographies' of flight and escape."[36] Returning to London from the Peace Congress in Paris, Brown receives an invitation to visit Hartwell House, the country estate of a Dr. Lee. In describing the carefully managed landscape, Brown takes pains to highlight its beauty. The first morning of his visit, Brown follows the gardener and discovers "the richest specimens of garden scenery" including "peaches hanging upon the trees that were fastened to the wall; vegetables, fruit, and flowers . . . in all their bloom and beauty; and even the variegated geranium of a warmer clime . . . there in its hothouse home, . . . seem[ing] to have forgotten that it was in a different country from its own" (90). Brown points to the intensive management of the garden—the espaliered fruit and temperature regulation of the hothouse—but does not associate these with the violence of British colonial plantations. Instead Brown emphasizes Lee's "taste in the management of his garden" and notes that he has seldom "seen a more splendid variety of fruits and flowers in the southern States of America" (90).

While as Schoolman notes Brown does increasingly linger on the "managerial repression and prematurely forgotten violence" of European colonialism as *Three Years* progresses, this description of the garden emphasizes the civility of English cultivation in contrast with American slavery, and it sets up for Brown to expound upon these differences the next day when he tours the cottages of the estate's laborers.[37] The laborers themselves are absent (presumably working) during this tour, but Brown describes their homes in idyllic terms. After moving through a tidy cottage, "a picture of neatness, order, and comfort," Brown describes the tasteful aesthetics of the front yard, including its flower beds and holly plant (92). The comparison culminates when Brown "call[s] the attention of one of my American friends to a beautiful rose near the door of the cot, and sa[ys] to him, 'The law that will protect that flower will also guard and protect the hand that planted it'" (93). The implication, of course, is that the same is not true in the United States. Brown's statement to his friend describes an American system in which plants are valued more than the enslaved who cultivate them. Using a rose, rather than a crop plant, to make this point draws attention to the great value placed on garden aesthetics and the kinds of power inherent in these types of display.

Northern Improvement

At the same time that abolitionists compared agricultural practices of the U.S. South with those elsewhere, horticultural developments in New England helped consolidate the power of northern farmers. Douglass's increased attention to the scientific expertise on the Lloyd plantation in his second autobiography and his interest in the D.C. landscape need to be understood alongside the years that he spent in Rochester, New York.

Douglass moved to Rochester in 1847 when the city was rapidly becoming the center of the American seed industry, and it would soon acquire its designation as the "Flower City" for its concentration of flower nurseries, seed shops, and horticultural periodicals. The city's proximity to various forms of transport—including boat traffic on Lake Ontario and the Erie Canal—made it uniquely situated for the distribution of plant materials to western states. Several major agricultural and horticultural treatises were written by horticulturalists living in the region, such as Patrick Barry's *The Fruit Garden*, printed by Scribner in 1851, which expounded on the "desire of every man, whatever may be his pursuit or condition in life, whether he live in town or country, to enjoy fine fruits, to provide them for his family, and, if possible, to cultivate the trees in his garden with his own hands."[38] When the Douglass family moved from the center of town to a farm on the city's outskirts that offered more privacy for relaying fugitive slaves north to Canada, Douglass tried his hand at peach cultivation.

During the middle decades of the nineteenth century Rochester also became a central site in efforts to organize farmers regionally and nationally. The South was not alone in seeking remediation for exhausted soil, and the agricultural reform movement of the mid-nineteenth century sought to create conditions conducive to farmers' prosperity in the North. Scientific agriculture was an important component of this movement. Seen as a boon to farmers, the practice of scientific agriculture promised to maximize profit, sustain yield, and make the lives of agriculturalists easier. It also helped cohere a dispersed population around a common cause. As Ariel Ron has shown, the agricultural reform movement of the antebellum period "aligned a substantial portion of the country's overwhelming rural majority behind a program of agricultural education, research, and development" and then "summoned the state to meet its needs."[39] As a result of periodicals and fairs, farmers became better networked and politically organized, exerting political influence over the creation of federal bureaucracies catering to their interests.

Agricultural "improvement," as it was deemed, might thus be another important avenue of what Britt Rusert has called "fugitive science," scientific practices and epistemologies that African American writers turned to in the struggle for freedom. Rusert has powerfully illustrated how nineteenth-century African American writers responded to the racial science of the day, particularly the ethnology that served as the basis for arguments of racial inferiority. But she also points to figures who were interested in how "sciences with no particular connection to the science of race could be used to enact a radical concept of freedom."[40] This is indeed the case with the rise of soil chemistry and other scientific practices with explicit applications for cultivation.

Agriculture was an important and politically charged field of scientific innovation in the nineteenth century, and it was linked to other fields of knowledge. As previous chapters have shown, the relationship between horticulture and agriculture was often thinly defined. Beth Tobin has broken down botanical science in late eighteenth-century England into four categories: economic botany, scientific botany, horticulture, and polite science.[41] Each category generally connotes different practitioners, from rugged men of science tramping through swampy terrain in search of profit on bioprospecting missions to genteel women botanizing in the home garden and parlor. But while the list gives a sense of the range of botanical practice, such categorizations can create false distinctions, making practices and knowledge that were fluid appear wholly discrete. For instance, agricultural and horticultural periodicals continually blurred any disciplinary line between the two, publishing on topics related to cultivation widely conceived. Within the same pages readers might encounter articles on potting flowers and learn about sourcing indigo. And journals that were not exclusively dedicated to agriculture and horticulture also regularly covered developments in both. African American periodicals in the antebellum and Civil War period routinely published articles related to horticulture, agriculture, and home gardening. The *Christian Recorder*, for instance, maintained several regular columns on "House, Farm, and Garden," "Science and Art," and "Popular Natural History." Published in Philadelphia, the *Recorder* had a national readership.[42] Likewise the *Colored American*, the *Provincial Freedman*, and the white-run abolitionist paper the *National Era* (which serialized Stowe's *Uncle Tom's Cabin*) routinely published articles on plant science and horticultural expertise.

As founder and editor, first of the *North Star* (1847–1851) and then of *Frederick Douglass' Paper* (1851–1863), Douglass regularly engaged with agricultural news, including the material published in local horticultural periodicals. In fact, one of the early financial sponsors of the *North*

Star was James Vick, who became one of the largest seed merchants in America. The *Genesee Farmer*, a popular and influential farming periodical, was also published out of Rochester in the mid-nineteenth century. Begun in 1831 by Luther Tucker, the periodical ran continuously during the years that Douglass was working as editor. In addition to the *Genesee Farmer*, Douglass regularly reprinted articles from the *Cultivator* (Albany, 1834–1865) and a number of other prominent farming periodicals. Such articles ranged over a wide array of topics, including everything from practical farming tips to depictions of foreign gardens.

Included in this range were articles that framed U.S. slavery in relation to global networks of capital. In 1848 Douglass reprinted a speech by British MP and antislavery activist George Thompson from the *London Mercury*. Thompson's focus is on expanding the trade relationship between England and India as a means of ending slavery and the slave trade in the Americas. Slavery began, he argues, "in a desire to obtain by forced labor the products of the earth."[43] Citing demand for sugar, coffee, cotton, rice, and tobacco as "the nourishment and vitality of these systems," Thompson blamed the British market's desire for raw cotton for the fact that slavery in the United States did not die out by the end of the eighteenth century. His argument steeped in imperial rhetoric about "this great empire" (4), Thompson called for a reorganization of international trade to shift cotton production to British India. "Slavery now lies entrenched behind its cotton bags . . . and the efforts of the British or even American abolitionists to dislodge it by moral suasion, we fear will prove . . . ineffectual" (17).

Skewering Carlyle's economic argument, Douglass likewise reprinted a blistering piece from *Punch* in response to Carlyle's essay, "Occasional Discourse on the Negro Question." The article satirically suggests that Carlyle was to be "rewarded by the West India planters for his late advocacy of 'beneficent whip,' and the Kentuckian wrath with which he has all but destroyed emancipated 'Black Quashee,' the wretch who will not work among sugar canes, unless well paid for his sweat, preferring to live upon pumpkin! To be, in fact, a free, luxurious citizen of accursed Pumpkindom."[44]

Douglass's papers provided a medium that could easily situate economic and moral suasion side by side. Scholars of nineteenth-century print culture have illustrated and contested the ways in which news-papers functioned as a medium for network building and connecting communities through shared readership and subscriptions.[45] As Benja-min Fagan has recently argued, the Black press played a pivotal role in antebellum Black activism, in terms of both content and the dynamics of the medium. And Eric Gardner's comprehensive study of the *Christian Recorder*, the African Methodist Episcopal Church's publication, under-scores both the broad reach of the newspaper as well as the ways that print was "far from the only joining force" in building the church com-munity.[46] As an editor, Douglass was well aware of the ways in which his paper might materially transform the lives of readers.

For instance, Douglass in the *North Star* was an advocate of Gerrit Smith's plan to redistribute 120,000 acres of his property in rural Upstate New York to African Americans.[47] Smith, a white social reformer, had inherited the land along with enormous wealth from his father, and he turned the practicalities of reallocating some of that land over to a team of prominent Black abolitionists, among them James McCune Smith, the physician and writer who wrote the introduction to *My Bondage and My Freedom*.[48] As Daegan Miller has argued, agrarianism had al-ways held radical potential as a means of establishing self-sufficiency, though southern planter culture had co-opted this ideal and distorted it into a vision that depended on slavery.[49] The Smith lands represented an opportunity to create an autonomous community of free Black prop-erty owners and to accrue wealth in the form of land. They also had the potential to contest southern agrarianism and model a different re-lationship to the land, what Miller has coined "eutopian agrarianism: the practice of making a place good."[50] But the difficulties of successful establishment were not unknown. In 1849, a year after the *North Star* ran its first notice of the upstate experiment, Douglass wrote an appeal to his readers for money to contribute to setup costs for the farmers, laying out the stakes: "These lands must soon be made a blessing to the colored people," he began, "or they will become a curse. We have hoped and expected much from them as a means of final independence to a large number of disenfranchised of this State."[51] Without money to get through the first season, the land's potential would be rendered worth-less. Pledging to contribute twenty-five dollars to the cause, Douglass encourages his readers to do the same, concluding the article by directing readers to send donations for this purpose to him and promising to acknowledge donors by name in the paper.

Douglass's own opinions on farming appear somewhat ambivalent, but he occasionally praised agriculture, along with the mechanical trades, as significant for social uplift. On September 22, 1848, Douglass reprinted in the *North Star* the committee address at the Colored National Convention in Cleveland, Ohio, from earlier that month. Decrying slavery in the South and identifying how "we are slaves of the community" in the North, Douglass and the other delegates of the National Convention identified labor as a crucial issue in the fight for racial justice. Urging their audience to move away from menial employment, the committee of delegates encouraged free Blacks in the North to "press into all the trades, professions and callings into which honorable white men press." Leaving aside the fact that many "fellow countrymen" were enslaved in agricultural labor in the South, the committee offered this directive: "Let us entreat you to turn your attention to agriculture. Go to farming. Be tillers of the soil. . . . Our cities are overrun with menial laborers, while the country is eloquently pleading for the hand of industry to till her soil, and reap the reward of honest labor." This call to farming reframes agriculture as a means of progress rather than a form of degradation.

A month later Douglass printed a short piece with the title "Farm Life" by the white abolitionist Horace Greeley. Greeley bemoans a common tendency among farmers' sons to reject the "drudgery and unintellectuality" of their fathers' vocation in favor of city life. Seeking to correct what he believes to be a grave error, Greeley urges that "in no vocation is scientific knowledge and mental elevation more desirable or more useful than in that of the farmer."[52] Cultivating the soil as a farmer, Greeley believed, corresponded with cultivation of the mind. And on March 18, 1853, another article appeared in *Frederick Douglass' Paper* with the title "Make Your Sons Mechanics and Farmers—Not Waiters Porters and Barbers."[53] Appealing to parents, the article urges readers to teach their sons a trade: "If *most* men will *not* teach our sable children trades in this country, some men *will*; and of the aid of the latter, we should quickly avail ourselves. There are mechanics and farmers, scattered over the country, who could be prevailed upon to take our sons into their workshops and upon their farms, and no colored man is excusable who does not seek out for his children such advantages." Trade is distinguished from these other occupations in that it does not involve waiting on others or waiting for work. Rather the work is characterized, the article continues, by industry, which might serve as a rebuttal to colonizationists. "Let the black man become an industrious and skillful mechanic, or a steady and industrious farmer, and there will be an end to all schemes for colonizing him, and all that wicked legislation designed to drive him out of the State. Let colonizationists bestow one tenth of the sum they expend in sending us out of the country in efforts to make us useful where we are, and they will soon find no motive for our removal elsewhere."

While Douglass printed such articles, he vacillated on the role of agriculture in Black uplift. In a letter to Harriet Beecher Stowe written in March 1853, Douglass praised the efforts of Smith and others to secure land for Black farmers in Upstate New York and Canada. Yet he expressed pessimism that it would make a difference. "Agricultural pursuits are not, as I think, suited to our condition. The reason of this is not to be found so much in the occupation, (for it is a noble and ennobling one,) as in the people themselves." Elaborating, Douglass emphasizes that systematic oppression has prohibited the majority of free Blacks from having the means "to set up for themselves, or get to where they could do so."[54] To overcome this difficulty, Douglass proposes the creation of an industrial college and asks Stowe to help with the practicalities of this endeavor. Such a college would "prepare men for the work of getting an honest living—not out of dishonest men—but out of an honest earth."[55] This appeal to an "honest earth" echoes Douglass's earlier description of the lessons to be gleaned from growing pumpkins in one's own garden.

Douglass's ambivalence about agriculture as a means of racial advancement plays into his developing views on the political possibilities of Western land. Douglass was bitterly disappointed at the racial restrictions of the proposed Homestead Bill that went before Congress in 1854. The bill, which promised public land in the West to settlers, contained a provision that would restrict application to whites only. Douglass penned an outraged editorial in response. Calling again on the neutrality of the land itself, he asks in perplexed outrage, "What kind of men are those who voted for the Homestead Bill with such an amendment! Do they eat bread afforded by our common mother earth?"[56] He continues by highlighting that many Black people were "Americans by birth . . . the first successful tillers of the soil," whereas many of the whites who would be able to claim land lack this association, being "foreigners, aliens, Irish, Dutch, English and French." The bill failed, but when the 1866 Homestead Act later explicitly encouraged African Americans to apply for land out West, Douglass became one of the most outspoken opponents of the idea.

If Douglass vacillated on the subject over the course of his career, his papers provide a window into the ways that a broader audience perceived the liberatory potential of scientific gardening and agriculture. One reader who advocated a more optimistic view in Douglass's paper was Pittsburgh minister and educator Lewis Woodson. An African American abolitionist who early on decried Garrison's methods, Woodson had been publicly chewed out by Douglass in the pages of the *Liberator* in 1850. But three years later, after Douglass's split with Garrison, his name appeared in an editorial in *Frederick Douglass' Paper* on the subject of scientific gardening. Woodson sums up his favorable impression of a recent trip to the Pittsburgh Horticultural Society exhibition and poses that such activity might be greatly beneficial to readers. Before enumerating the benefits of gardening, Woodson emphasizes the importance of wealth. "Mr. Editor," he begins, "Among the various means of our elevation and enfranchisement, I know of none more efficient than wealth, I am aware that learning is before wealth, but of what advantage is learning without the means of making good use of it?"[57] Wealth, Woodson continues, translates into power, and "if we are wealthy and respectable, wealthy and powerful, nobody would think of revolting at our leadership or of putting us down, for the very simple reason that they couldn't do it."

What this has to do with gardening becomes clear in the next paragraph. Woodson describes the rapid expansion of the Pittsburgh horticultural society in the past ten years, and his experience attending the recent exhibition where the "vast saloon was covered with the plants, shrubs, vines, and flowers of almost every mountain, hill, valley, and plain, on the face of the globe." Noting the splendor and the diversity of the plants on display—including an "enormous Peruvian pumpkin, almost the size of a common flower barrel" and gigantic watermelons—Woodson makes the point that none of these fruits or vegetables existed before the recent arrival of scientific gardening, given the "rough, rocky, cold nature of the soil, and changeable climate." Moreover, he cites initial skepticism that gardeners would be able to overcome these environmental conditions even with the adoption of new techniques.

But Woodson's conclusions extend beyond his assessment of the remarkable horticultural displays to the practical benefits that gardening might have for readers. Calling gardening a "healthy, invigorating business" as well as a virtuous one, Woodson emphasizes its viability as an easy and profitable business. He cites the story of one mechanic who struggled to make ends meet in the city but who thrived when he and his family moved out of town and planted a small garden. "How can gardeners fail to make money?" he asks rhetorically near the end of the article. Profit is all but assured, Woodson continues, and such employment answers Horace Greeley's call to "do something."

Several years later an article from "Our New Haven Correspondent" appeared in the pages of *Frederick Douglass' Paper*, "continuing our notices of the signs of the times, as they appear in this part of the field of vision." Characterizing the city's slow "change in public feeling," the author, signed A.G.B. (likely the African American pastor and activist Amos Gerry Beman), hailed a horticultural award granted to a local Black farmer as a sign of progress. "The New Haven County Horticultural Fair has just closed its annual meeting, and we are gratified to note the fact, that the second prize for the best squashes was given to Mr. M. Lyman, a colored farmer of the city. Let this fact stimulate others to 'go and do likewise.'"[58]

Lyman's award-winning squash was, by all accounts, fairly exceptional given the way that racial prejudice limited the participation of African American farmers at state fairs. The Toronto-based *Provincial Freeman* on October 28, 1854, described, for instance, how a Black man had submitted "some very fine grapes" at a Horticultural Exhibition in Philadelphia a few years earlier, which everyone said "were the best there." However, the judge decided not to grant an award for grapes that year, citing for a reason that "the season had been so wet." This example comes in response to an article reprinted from the *Ed. Standard* that described a recent "baby show" at the Ohio State Agricultural Fair (a harbinger of Fitter Family contests) where "one hundred and twenty-seven babies were offered as candidates for valuable prizes." The article ends by registering that Lucretia Mott had protested that only white babies were admitted. Picking up on this thread, a writer notes that had Black babies been admitted, they would have been ignored, for "when the prejudice extends to fruits and flowers, it is not probably that black babies would be viewed except as the subjects for mirth."[59]

By the mid-nineteenth century agricultural and educational reformers, as their periodicals demonstrated, were largely in agreement about the desire for scientific education for farmers and gardeners. Suggesting that the same skills might apply in both contexts, an article in the *Albany Cultivator for April*, reprinted in the *Colored American* in July 1838, opened with the claim that "every farmer should have a garden, for health, for pleasure, and for profit."[60] After elaborating on each of these qualities, the article promises the future publication of "the scientific principles of gardening of Prof. Rennie . . . [which] will be found no less applicable to the farm than to the garden." In July 1839 the *Colored American* reprinted an article by Josiah Holbrook about the agricultural education of farmers from the *Religious Telegraph and Observer*. He begins his letter to the editor by stressing his belief in the need for scientific education for farmers: "Mr. Editor,—I have for several years, been fully convinced, that neither lawyers, nor physicians, nor clergymen, nor professors of colleges, nor any other class of the community, have so many inducements or so many facilities for becoming really intelligent, scientific men, as farmers." Holbrook enumerates the range of subjects that a farmer daily encounters, including "botany, mineralogy, geology, chemistry, natural philosophy, entomology, and the natural history of animals." This familiarity through the everyday tasks of farming breeds expertise. Holbrook rhetorically asks, "Cannot almost any farmer give the best scientific botanist much useful information about plants?" This comes in part from daily use and in part from having farm fields in which to experiment: "A farm is a far better place for acquiring really useful knowledge, and for acquiring it more thoroughly than any hall of science, which is or can be constructed and furnished by the hands of men. It is a '*Cabinet of Nature*,' more richly furnished with specimens, and a laboratory where chemical and philosophical experiments are going on upon a larger scale, than can be found in any High School,

Academy, or College." A year later the newspaper reprinted an article on the topic of "Science Applied to Agriculture."[61] The article begins by attributing every material fact of life to the operations and laws of chemistry including "the clayey crucible" of the body. Analogizing the world as a whole to "a grand laboratory, where chemical action is continually going on" and describing nature as "a great workshop, where she is continually carrying forward her operations," the author makes chemical principles and laws essential to the farmer as a means of establishing a holistic balance. "To imitate and assist her in carrying this law into effect, is part of the service of the farmer, and in proportion as he does his duty, will his labors be rewarded."

If reward in the form of recognition was sparse for African American farmers at state agricultural fairs, efforts existed to create Black fairs at the regional level. Douglass was one of the speakers at the Colored People's Agricultural Fair at Geneva, and the fair was written up in *Frederick Douglass' Paper* in October 1854 with decidedly mixed reviews. Noting that it was "chiefly interesting, as being the first attempt of the colored people in that section to prove their skill and industry by exhibiting their productions," the article observes that it was "very small indeed." Anxious for a better future turnout, the article lays out the stakes: "We do sincerely hope for the credit of the colored people, that in future when there is an exhibition of this sort, the colored people generally will pull together and bring out their resources. . . . Its successors, to save us from ridicule and contempt, must be a decided improvement on the first. It is fashionable for colored editors and writers to puff anything and everything of a colored origin. We do not follow that fashion."[62] Appearing in the paper in June of the following year was another short unauthored essay distinguishing between book learning and practical experience when it comes to farming. "He who is a good theoretical agriculturalist may be a very poor practical farmer. The two kinds of knowledge, theoretical and practical, require, to a certain extent, a somewhat different order of mind."[63]

The articles that Douglass chose to reprint from these periodicals demonstrate not only the range of coverage in agricultural periodicals but occasionally their political commitments. Farmers and politicians alike used newspapers to advocate state and federal policies. One such article, reprinted in *Frederick Douglass' Paper* on February 19, 1852, and attributed to the *Albany Cultivator,* offered a description of the Jardin des Plantes in Paris that makes the case for scientific education. Encouraging the United States to adopt better public scientific education, the article commends the famous Parisian gardens as a public institution that educates the public about natural history and other scientific fields. The garden contains "an extraordinary collection of the natural productions of the earth, including besides plants and forest trees, animals from all quarters of the globe, birds, minerals, collections in comparative anatomy, and indeed every department of natural science." Leaving aside the imperial nature of collecting plants, trees, and animals from "all quarters of the globe," the author emphasizes the democratic nature of the collections, and moreover notes that "there is an amphitheater with a laboratory and apparatus for public lectures, which are given gratuitously on every branch of science for more than half the year." Holding up the Jardin des Plantes as a model, the article contrasts the Parisian institution with the state of scientific education in the United States. The United States lacks resources for such facilities, the author continues, because it wastes money on violence against Native nations: "The money expended freely and willingly by the United States, in driving out or exterminating a single tribe of Indians, would set up half a dozen such institutions as the Garden of Plants at Paris."[64] With disdain for federal Indian policy, the author pits bad policy against scientific progress. The argument is fundamentally economic: whereas Native American removal policies were a money sink, scientific education held the promise of democratizing scientific discoveries that might yield economic growth.

Douglass's paper afforded an important medium for disseminating perspectives that linked such scientific progress—or lack thereof—to political climate. Given the importance of agriculture to many African Americans after the war, particularly in the South, Douglass clearly paid attention to and promoted its possibilities for social and economic advancement, presenting various threads on the subject over the course of his editorship of two papers. Whereas scientific agriculture grew popular among the planter class as a way to sustain slavery, Douglass and those who wrote for his paper recognized its potential for Black economic mobility, and the way it might be leveraged against bad politics.

* * *

In September 1873 Frederick Douglass stood in front of a crowd of freed-men at the third annual Tennessee Colored Agricultural and Mechanical Fair to speak "of the importance of agricultural and mechanical industry and of united effort on the part of our people to improve their physical, moral, and social condition."[65] Addressing a fairground of several thousand in Nashville, Douglass began his speech with the admission that he actually knew very little about the occupations of his audience:

> Gentlemen, this is an agricultural and mechanical industrial fair. I am surrounded to-day by industrious mechanics and farmers, and you have got me up here to tell you what I know about farming. Now, I am neither a farmer nor a mechanic. During the last thirty-five years I have been actively employed in a work which left me no time to study either the theory or practice of farming. I could far more easily tell you what I don't know about farming than what I do know, though the former would take more time to tell than the latter. (5)

Douglass's admitted lack of knowledge about the theory or practice of farming corresponds with the Douglass who tends to be remembered today. In each of his three autobiographies and across many of his speeches, Douglass aligned his own personal development with his escape from slavery and rise to prominence within the abolitionist movement. This trajectory is a movement away from the plantation experiences of his early life. Douglass's emphasis on writing, editing, and oratory as tools of social reform were the focus of his life's work. Standing before this crowd, Douglass confesses his limited understanding of their circumstances and acknowledges the ways in which his own life has not been representative of theirs.

Yet in a number of critical ways Douglass spoke of agriculture both for its role in oppression and a means of potential racial uplift. In Tennessee in 1873, agricultural land and labor were key issues in the reconstruction of society. While vehement white opposition had checked the Freedmen's Bureau's power to redistribute land to Black people, a small number of Black farmers did manage to purchase land during Reconstruction and nonlandowning Black farmers negotiated for more autonomy.[66] For these farmers, conditions of working the land were tied to the question of economic prosperity. As he continued his speech in Nashville, Douglass noted that "the commanding question of the hour" is how agriculture "can be made to serve us, as a particular class" (10). For Douglass, the answer lies partially in the collective formed around this enterprise. Douglass praised the crowd, noting, "You have wisely availed yourselves of a well-known power, the power of association, organization, mutual counsel and cooperation. You have dared to organize an Agricultural and Mechanical Association for the State of Tennessee. . . . You have dared to open here, in the city of Nashville, a State Agricultural Fair, to display the rich fruits of your industry, and to ask your fellow-countrymen, of all conditions and colors, to view and inspect them. This is an act, on your part, as brave as it is wise" (379). Association, organization, mutual counsel and cooperation: these are building blocks for political power. Douglass praises this activity as cohering around particularly modern developments, noting the vast scientific and technological improvements that had revolutionized agriculture across the middle decades of the century. Modernity, on this view, is a positivist phenomenon. "Science," Douglass proclaimed, "the noblest and grandest artificer of human fortune and well-being, the source and explanation of all progress, has patiently unfolded the nature and composition of plants, and made us acquainted with the properties of the common earth, wherever they grow" (380). Soil and weather can be broken down into chemistry and "irreversible laws" (381). Fields may be rationally, scientifically planted. Which is not to say they can be farmed with detachment.

One remarkably consistent dimension of capitalist markets is the ways in which exhaustion of a resource rarely signals the end of exploitation. As Stephanie LeMenager demonstrates in her landmark *Living Oil*, the end of "easy oil" has not meant large-scale adaptation to living with less; we remain besotted by petroleum and the lifestyles it enables.[67] Likewise, exhausted southern soil signaled not efforts to scale back production or the end of planter wealth, but a problem that required adaptation. And as LeMenager points out, adaptations in response to limited resources are rarely less extractive and often much more destructive. Planters moved West when soil failed to respond, or they looked to chemistry and soil amendments to remediate the exhausted land, rapidly expanding the fertilizer market. As Jennifer James notes, guano teaches us a lesson about how capitalism disavows its own destructiveness: "It damages and then endeavors to cover that damage by marketing us a solution."[68] These solutions might prop up a system that is mutually exploitative of humans and earth, and sustain the system for another season, and yet another.

Douglass makes a crucial distinction between plantation agriculture and small-scale farming. Talking about the ways that slavery governed relationships between humans and the land, Douglass reflects that "the very soil of your State was cursed with a burning sense of injustice. Slavery was the parent of anger and hate. Your fields could not be lovingly planted nor faithfully cultivated in its presence. The eye of the overseer could not be everywhere and cornhills could be covered with clods in preference to soft and pulverized soil in their absence, for the hand that planted cared nothing for the harvest. Thus you will see that emancipation has liberated the land as well as the people" (12). Depleting the soil through monocropping is not the only means by which the land suffers. Douglass critiques slavery here for the way that it forecloses care in the relationship between field workers and the landscape.

Going beyond a utilitarian argument against large-scale agriculture, Douglass establishes the possibility of an affective land ethic. The landscape is entirely transformed not just by the size and scale of the farming enterprise, but by the related exploitation of people and soil. The key words here are the adverbs that characterize planting and cultivating: *lovingly* planted, *faithfully* cultivated. Slavery precludes an affective investment in the land when "the hand that planted cared nothing for the harvest." Scholars have long recognized the ways that Douglass emphasizes slavery's rupture of familial relations.[69] Here he extends this critique to encompass an affective relationship between the enslaved and the environment. The argument here extends beyond characterizing slavery's dehumanizing effects to articulate its harm on the landscape that seems divorced from economic calculations. Implicit in Douglass's claim is a providential ethos: if emancipation liberates land as well as people, it opens space for continued cultivation that is rooted in care. The land may remain under the plow, but with a crucial difference of the way that the farmer drives the plow.

Cultivation in this light is inextricably bound to issues of slavery and freedom, exposing the limits of a horticultural rhetoric that does not take this history into account. Douglass too gardened as a hobby, and often set this activity apart from his activist work. While his tireless travel schedule during his busiest years of activism often kept him away from home, friends and family remember the gardens that Anna Murray Douglass kept at each of their houses, and Douglass maintained his interest in tree culture once the family moved to Cedar Hill, the home in Anacostia outside of D.C. where he lived at the end of his life. By many accounts, Douglass devoted increasing time to gardening and horticulture at Cedar Hill. Ka'mal McClarin, curator at the Frederick Douglass National Historic Site, has documented that Douglass not only gardened but was a "conscious collector of different types of flora and fauna" and pressed floral specimens into books in his study. The list of specimens, as McClarin notes, was wide-ranging: "mustard, lesser club moss, wild rose, crimson clover, jasmine, wood sorrel, grape hyacinth, Kentucky bluegrass, thorny bamboo, sweet gum, red maple, trumpet honeysuckle, and ferns."[70] And Douglass's outdoor retreat at Cedar Hill, named the "growlery," a Dickensian neologism from *Bleak House,* was perched at the edge of the estate's orchard.

That Douglass could tend strawberries at Cedar Hill late in his life without recourse to the market is a privilege that separated him from many in his Tennessee audience. Douglass's position illustrates the tensions inherent in horticultural discourse as well as in his own self-presentation as a representative individual, particularly with the rise in his class status. Scientific gardening and agriculture were never at the center of Douglass's fight for racial justice, but they were significant in his thinking nonetheless given their pervasive cultural influence. Across his career Douglass navigated the material and metaphorical meanings of cultivation, and the fault lines they produce caution against a neat conceptual understanding of the garden politic.

Conclusion

An Ethos of Collectivity

Nineteenth-century Americans' immediate, intimate experiences cultivating plants brought them into a caretaking relationship with forms of life that were organized differently from their own. They reflected seriously on the significance of plant life for their own lives, and in paying close attention, many saw models of collectives and of dynamic multispecies interactions that escaped taxonomy's ready vocabulary. Gardeners were drawn to care practices that stood in sharp contrast with the extractive practices typical of the contemporary political ecologies they otherwise saw all around them.

Today we live with the legacies of these dominant political ecologies and have inherited their ways of seeing. The long arc of colonial bioprospecting has made it conceptually difficult to embrace plants as complex beings today, let alone to recognize a longer genealogy of thought about plant vitality. Yet over the past decade, the incredible capacities of plant life have gained more attention in popular culture, and artists and scientists have sought to illustrate how an affective and intellectual appreciation for plant life might expand the horizons for political care. These efforts resonate with the work of nineteenth-century authors, scientists, and cultural theorists who sought to use their understandings of plants to think through the contours of an exclusive polity.

That this earlier work has been largely forgotten underscores the force with which imperial science and capitalism have dominated environmental perceptions and realities. Plants are still "big science and big business" in contemporary Western culture, treated as valuable objects available to human interests rather than sensitive subjects and agents in their own rights.[1] Drug companies continue to comb the rainforests for medicinally valuable species, trees are grown to harvest for timber, and crops are intensively managed through chemical applications. Anthropogenic climate change wreaks havoc on habitats, and whole eco-systems are changing as new weather patterns shift blooming dates and seasonal cycles, forcing plants to adapt or migrate. As "perpetual snow" becomes a polar rarity, the plants on Humboldt's Naturgemalde inch up the mountain. Many do not survive. Those that do are subject to shifting regulation when they migrate across political borders.[2] For many Indigenous communities, such migrations have already amplified the ongoing legacy of colonial violence.[3]

Mainstream U.S. culture has made it easier to imagine plants as passive recipients of our desires. As the Potawatomi scientist and writer Robin Wall Kimmerer has noted, the English language itself makes it easy to flatten the liveliness of other forms of life. "English doesn't give us many tools for incorporating respect for animacy," she writes. "In English, you are either a human or a thing. Our grammar boxes us in by the choice of reducing a nonhuman being to an *it*, or it must be gendered, inappropriately, as a *he* or a *she*." By contrast, in Potawatomi plants and mountains and bodies of water are all alive. The difference has a profound impact on the way we value the world around us because grammar maps relationships. "Maybe a grammar of animacy could lead us to whole new ways of living in the world," Kimmerer posits, "other species a sovereign people, a world with a democracy of species, not a tyranny of one—with moral responsibility to water and wolves, and with a legal system that recognizes the standing of other species."[4]

Kimmerer writes with a sense of conviction that a great shift in collective consciousness is possible. And there are encouraging signs that she is speaking to an increasingly receptive audience at the intersection of popular culture, science, and anthropology. Numerous recent books, articles, and podcasts on the subject have illustrated the keenness of readers and listeners to embrace these ideas. The popular science show *Radiolab* has covered plant intelligence. David George Haskell's Pulitzer Prize finalist for nonfiction, *The Forest Unseen* (2013), describes plants' capacities for behaviors that surprise us. Michael Pollan's provocatively titled "The Intelligent Plant: Scientists Debate a New Way of Understanding Flora" caused a stir when it first appeared in 2013 in the *New Yorker*. Offering a detailed view of the quarrel in scientific communities over the contested subdiscipline of "plant neurobiology," Pollan writes that some proponents argue we must "stop regarding plants as passive objects—the mute, immobile furniture of our world—and begin to treat them as protagonists in their own dramas, highly skilled in the ways of contending in nature."[5]

The new field's detractors in the plant science community take umbrage at the use of the term "neuro" to define something without a brain. And the stakes in the debate, Pollan points out, are high: "This issue generates strong feelings, perhaps because it smudges the sharp line separating the animal kingdom from the plant kingdom. The controversy is less about the remarkable discoveries of recent plant science than about how to interpret and name them: whether behaviors observed in plants which look very much like learning, memory, decision-making, and intelligence deserve to be called by those terms or whether those words should be reserved exclusively for creatures with brains."[6] The contest over semantics indicates some real investments in maintaining a clear boundary between plants and humans even now. The idea of plant consciousness raises the ethical implications of conceding plant intelligence, or plant pain.[7] To some the comparison is problematic because it elevates plants by making them appear more like us. But for others, characterizing other-than-human life using terms like "consciousness" actually occludes the intricacy and distinctiveness of plant life. Anthropologist Natasha Myers has underscored that to many, though not all, plant biologists, articles like Pollan's seem like polemics that recklessly anthropomorphize plants, making them seem lesser by comparison to our own forms of feeling and cognition and inhibiting us from recognizing their unique capabilities.[8] The language of comparison, in short, delimits our imagination of this other form of life, rather than unfurling our imaginations toward radical alterity that lies at or beyond our comprehension. Geobiologist Hope Jahren ends her 2016 memoir *Lab Girl* by stating plainly that "plants are not like us" and asks her readers to instead appreciate their "deep otherness."[9] That is, she is asking readers to imagine how they exceed our capacity of understanding and to modify our behavior so as to respect that life that lies beyond our comprehension.

But the comparisons between humans and plants continue to beguile, no doubt because such identifications often lead toward profound shifts in worldview. In 2015 German forester Peter Wohlleben published *The Hidden Life of Trees* (2015), which quickly became a best seller and makes the argument that forest trees are communal, communicative, and even, at times, compassionate. Wohlleben shows how trees take care of each other and protect the most vulnerable, just as human political systems might do. He compares the ways that trees redistribute sugar to help neighbors to "the way social security systems operate to ensure individual members of society don't fall too far behind."[10] And he describes trees in human and domestic terms: "When you know that trees experience pain and have memories and that tree parents live together with their children, then you can no longer just chop them down and disrupt their lives with large machines."[11] Knowledge that trees have families and lives, he hopes, might shape human behavior.

Despite their opposing stances, Jahren and Wohlleben are aiming toward a similar goal. Underscoring difference and drawing attention to similarities are two different ways of paying close attention to plant life and making a case for better treatment. Both Jahren and Wohlleben work toward convincing a skeptical public that plants should challenge our mostly anthropocentric ethical norms. They are part of what Myers has characterized as a "plant turn" since 2010 "among philosophers, anthropologists, popular science writers" and their readers (a list to which we should add literary critics). The recent spate of books and articles on the science of plant communication not only present the complexity and sophistication of plant life but also ask how this might shift our understanding of ourselves, our communities, and our behaviors. The forest ecologist Suzanne Simard, for instance, has been working for decades to understand the role of mycorrhizal networks, involving symbiosis between fungi and plants, in plant communication. Her work has been translated by fiction and nonfiction writers into narratives of ethical realignment in which individuals in Western societies come to think more communally and altruistically.[12] A 2020 New York Times article described how Simard's research invites us to imagine how "an old-growth forest is neither an assemblage of stoic organisms tolerating one another's presence nor a merciless battle royale: It's a vast, ancient and intricate society. There's conflict in a forest, but there is also negotiation, reciprocity and perhaps even selflessness." Understood better as assemblages than individuals, forest plants might inspire a model for collective responsibility and resource distribution with implications that can radically reframe the history of life. As journalist Ferris Jabr stresses, this provides an alternative model of cooperation to the model of competition that Darwin and others advanced in the second half of the nineteenth century.[13]

Many of the authors who translate these scientific ideas for popular audiences do so as an expression of desire for political change and imagine that political change emerging through parallel change in the way that we collectively think about the species and ecosystems around us. Novelist Richard Powers also drew on Simard's work, and even included a character based on her, in his 2018 Pulitzer Prize–winning novel *The Overstory*. Set primarily in the last few decades of the twentieth century, the multiplotted novel covers a sprawling cast of characters whose individual lives unfold in the shadow of the catastrophic destruction of redwood trees. But it also makes clear that this destruction isn't the only thing to dwarf their perspective; the liveliness of massive, old-growth forests prompts many of the characters to sacrifice their time and even their lives to saving them. These characters see what many others do not, namely, how "*life runs alongside them. . . . Creating the soil. Cycling water. Trading in nutrients. Making weather. Building atmosphere. Feeding and curing and sheltering more kinds of creatures than people know how to count.*"[14] Much like Dickinson's use of verbs in "Bloom—Is Result—to Meet a Flower," Powers turns to gerunds to stress the agency of trees: they are continually and simultaneously creating, cycling, trading, making, building, feeding, curing, sheltering. These activities are as profound as they are invisible. Whole lives unfold at paces and timescales that frequently escape notice.

Powers believes that by paying attention to these lives we might choose to live our own differently. And indeed, he changed his own life after researching the book, moving from Palo Alto to the Smoky Mountains. In an interview with Everett Hamner in the *Los Angeles Review of Books*, Powers emphasizes our "great errors in thinking" that make human exceptionalism possible and praises multispecies narratives, stories where "people must lose themselves and their private narratives in an unseen network of connections that runs far beyond their own small selves, even beyond their own species."[15] The novel is an attempt to produce a collective shift in consciousness among his readers.

Students in my environmental literature classes tend to be captivated by such ideas but then wonder what, if we seriously take these ideas on board, we are going to do about it. Many are skeptical that such ideas are powerful enough to change behavior. Others worry that lending such agency to plants elevates another species while ignoring systems of violence and inequality within our own.[16] The popularity of plants in nineteenth-century America and their complex role in people's lives show that there is an unacknowledged precedent that offers a longer genealogy for thinking through these questions. What lessons might this earlier period and its rich literary archive offer for the contemporary plant turn?

First, nineteenth-century literature helps us identify the domestic sphere as an important vantage point for understanding environmental change. By looking to descriptions of plants in and around the home, we find forms of everyday environmental relation worthy of our attention. These narratives attend to what it feels like to live with plants—the ordinary, habitual, and extraordinary dynamics that unfold in daily relation and the structures of feeling born out of these dynamics. As plants became household objects, their meanings proliferated. Gardeners formed attachments to the plants in their care, affinities that transformed the way they imagined their relationships to those plants. Recovering these kinds of local knowledges can illustrate the divergent ends to which different actors mobilized acts of cultivation, especially as these actors sought to imagine different political horizons.

Moreover, while vegetal forms have always shaped linguistic expression, nineteenth-century literature's conspicuous attention to plant life very often pointed to the limits of language itself to attend to other forms of life. Mastery is one very consequential narrative of nineteenth-century plant circulation, but it is not the only one. Writers who attended to plants through study or care encountered resistance, frustration, and uncertainty. Rather than revealing the inner truth of plants, their writing often showed where language fell short. For many of the authors here, the inability to fully comprehend these other forms of life, the inability to capture that life in language, fascinated and humbled, prompting writers to reassess common paradigms of comprehension.

Crucially, while plumbing the material contours of plant life, these writers also expanded the metaphorical significance of plants. Elaine Freedgood has shown how we might read literary objects metonymically, tracing broader structures of production through specific purchases.

The writers discussed in this book show how structures of relation in the domestic garden—the pastoral, say, or the sentimental—were tied to the role of plants in the nation's political economy, to the geopolitics of bioprospecting and the environmental effects of empire. That is, they help us see how a popular domestic floral lexicon relates to the botany of empire, and how the former frequently occluded the visibility of plant life in the latter.

These earlier writers model the political possibilities that open up if plants' liveliness is taken seriously. Today those who compare plants to humans are frequently excoriated for anthropomorphizing. But writers living with plants in the nineteenth century both hailed the alterity of plants and sought to imagine how this other life form might instruct human society. One of the reasons why plant communication is such a radical contemporary idea is that it makes nature a realm of the collective rather than the individual. Biology for over a century has been driven by the primacy of the competitive individual or species. Lynn Margulis's late twentieth-century work to convey the evolutionary importance of cooperation was frequently attacked, dismissed, and treated as a fringe idea for many years before the scientific community started to accept her conclusions.[17] But the benefits of mutual aid to organisms were apparent even to Darwin. Indeed, as recent popular science writers have noted, plant "altruism" posed a problem to Darwin's theory of natural selection.[18] Contemporary writers are drawn to such ideas that change the way we imagine "nature," but the distributed way agency works in Powers's novel was already explored by Dickinson in her poetry over a century and a half ago.

The idea of distributed agency is alien to notions of representation that rely upon the idea of the bounded individual subject. Taking plants seriously on their own terms invites an ethical and political recalibration because it effectively disrupts common assumptions about the way that the world works. Philosopher Emanuele Coccia has stressed how narrowly Western culture has defined "life" in the past century, a fact that has curtailed our sense of responsibility toward plants and other fundamentally different life forms. Understanding plants as passive and inert allows us to use them as we wish and to make their meaning a product of our own desires. Granting them consciousness, by contrast, creates space for their autonomy, and potentially their rights.[19] It certainly troubles ingrained assumptions about the greenery around us. Daegan Miller notes the ways in which particular species in the nineteenth century had strong cultural valences, writing that "many in the nineteenth-century U.S. understood the distinction between, say, redwood and cedar as not just physical, but cultural: the two trees simply meant different things."[20] This might seem strange today because contemporary U.S. culture has largely lost such distinctions. Few are plant-literate to the point where they can identify the species they encounter in their daily lives. This lack of nuance speaks to a broader inattention and allows ready acquiescence to the great extinction events happening every day.

In a moment when environmental crisis poses a threat to democratic futures around the globe, paying attention to this earlier period's plant life helps clarify the political possibilities that emerge from reimagined relationships to other species. But if this earlier period teaches us to better see the stakes of the critical plant turn, we might do well, as Douglass reminds us, to be cautious about the potential for these ideas to transform our political worlds for the better. Plants in the nineteenth century were cherished home companions, scientific specimens, and crops. Such radically different yet fundamentally related contexts drive home the way metaphoric possibilities are governed by the uneven valuation of human life. As contemporary poet Noor Hindi writes, "Colonizers write about flowers. / I tell you about children throwing rocks at Israeli tanks / seconds before becoming daisies."[21] Focusing on how plants are like "us" might unsettle assumptions that help perpetuate political capitulation to extractive capitalism. And it might upend stable ideas about who or what deserves ethical treatment. At the same time, considering the liveliness of plants in relation to human qualities has the potential to consolidate an idea of the human, flattening difference and the ways that political policies exacerbate inequalities among human beings.[22] It can draw our attention away from pressing issues of social justice in human communities. White men in the eighteenth and nineteenth centuries made the status of humanity consonant with being white and male, and a universalizing impulse remains alive and well today. And racism has pervaded the mainstream environmental movement through every stage in its history. Over the past century many famous conservationists have prioritized the "protection" of stolen land in an illusory wilderness state over the livelihoods or rights of displaced populations. And even today the Western scientific community receives most of the credit for "discovering" ideas of plant animacy and communication. Kimmerer describes the ways in which animacy is built into Potawatomi language and cosmologies, but the turn toward understanding plants as animate and intelligent in the twenty-first century is often presented as the result of novel Western science. Almost every mainstream account of plant intelligence or communication reaches only as far back as the 1970s and then mainly attributes the ideas to a small number of eccentric white men. This woeful inaccuracy is part of a broader pattern in U.S. scientific thought going back to the earliest days of the republic. The established scientific community in the nineteenth century seldom acknowledged Indigenous forms of scientific knowledge (even as it drew on this knowledge) and worked to entrench Western empirical understandings of plants.

In this context, the value of the plant turn might simply be to draw attention to the ways that valuations of human life have been forged in historic interspecies contexts. Such consciousness raising that allows for the possibility of other modes of relation. One of the most important points that Kimmerer makes in *Braiding Sweetgrass* is that to recognize kinship with plants is to recognize all kinds of relationships with the more-than-human world and to proceed with reciprocity and gratitude. Paying attention to plants in this way does not mean elevating them to the exclusion of other beings. Rather, it is a mode of noticing how broad and deep run our debts to other forms of life, and changing our own lives to acknowledge this as responsibility. "One of our responsibilities as human people is to find ways to enter into reciprocity with the more-than-human world," Kimmerer writes. "We can do it through gratitude, through ceremony, through land stewardship, science, art, and in everyday acts of practical reverence."[23] Plants teach how interconnected everyone and everything is, so that learning their lessons might mean changing everything about our current systems of exploitation.[24]

In 2020, when COVID-19 shut down a number of workplaces and rerouted others through the home, it strained the set of ordinary and diverse practices that make up domestic life.[25] Home became the locus for many of the activities that used to happen in schools, workplaces, and spaces of leisure. The crisis placed an overwhelming burden on the housing-insecure, survivors of ongoing domestic abuse, parents, and especially single parents, those deemed "essential workers" and countless others already living in great precarity.

The seismic recalibrations of the domestic sphere driven by the pandemic also threw into relief the home as a site of environmental thought and practice. Heating and cooling and water bills went up, consumption patterns shifted, and houseplant sales accelerated across the nation in the months after the United States entered lockdown.[26] For many, nurturing plants provided a way to nourish hope in the face of so much hardship. People turned anew to seed catalogues and gardening centers, they laid out container gardens, and they paid increased attention to the growth in their yards. They also participated in others' domestic spheres remotely through sites like Instagram, Twitter, Facebook and TikTok, amplifying an already-ongoing social media conversation about plants and self-care.

Many of these conversations hinge on the premise that living with plants—being around and attentive to them in kitchens and bathrooms, entryways, and bedrooms—is important to the well-being of the human caregiver. Christopher Griffin, assistant director of NYU's LGBTQ+ Center and creator of the delightful and popular Instagram handle @ plantqween, has described how taking care of over 160 houseplants in their apartment has allowed for introspection and growth: "My green gurls continue to feed me that energy to carve the mental space to be reflective and introspective in my process. . . . Putting that love, care and attention into my plants has honestly provided me with practice to be able to put that same amount of love, care and attention into my own Black queer femme being."[27] During the height of the Black Lives Matter protests in 2020, the artist DJ Freedem drew on the ways that plants can be a basis for care practices, and also highlighted their continuing potency as political objects, when he posted on Instagram calling for white people to give plants to Black people. Freedem's call for plant reparations quickly went viral, outgrowing Instagram and leading to the launch of a website called the Underground Plant Trade, where the project is described as a combination of "community building . . . resource sharing . . . [and] satirical political commentary."[28] While the project highlights the urgent need for formal monetary reparations, this informal network also draws attention to the role of plants in well-being and world building.

Most fundamentally, the Underground Plant Trade acknowledges plants' implication in systems of power. On the website's description of the project, Freedem links the project to the rise of antiforaging laws in the decades following Emancipation, part of a broad pattern of increased governmental regulation that disproportionately targeted Indigenous and Black foodways and cultural practices. One legacy of the late nineteenth century is the failure of Reconstruction and the rise of Jim Crow. Another is the increasing federal regulation of plant and animal life through various forms of land management. As Freedem makes clear, these projects were intrinsically linked.[29]

Our yards and greenways tell a material story of the ways that nineteenth-century plant circulation reshaped both natural and political environments. Nineteenth-century botanical science helped forge botanical nationalism, solidifying a settler-colonial understanding of the environment.[30] Bioprospecting and monoculture laid the groundwork for some of today's most severe forms of injustice. Our food and commodity systems are the legacy of plantation systems, which consolidated profits through the violent coercion of labor. And the search for plant-based medicines today continues the process of bioprospecting.[31]

Many works in the environmental humanities have exhorted the need for new narratives, or the recovery of older narratives, that help make different patterns of relation possible. Narratives structure our imagination, and imagination, as Jedidiah Purdy reminds us, is often "intensely practical."[32] The most prevalent stories today are those that uphold current systems of extraction, but they seem increasingly open to imaginative challenge.

The need for political action on environmental issues that stem from nineteenth-century legacies has never been more apparent, as government policies and corporate activity exacerbate environmental degradation and social crisis. Nineteenth-century writers remind us that ideas of plant life have long been tied to ideations of the American republic. Reading their work can encourage us to recognize how fundamentally the political economy of plants structures our systems of power, and how alternative political visions can emerge when we pay serious attention to other forms of life, especially those that resist our comprehension and control.

ACKNOWLEDGMENTS

As a kid I spent as much time as I could alone in the small wetland that separated my house from the playing fields of the regional high school. Less than a half acre in size, this patch of ferns and skunk cabbage belonged to our neighbors Eddie and Peggy. My parents had bought a small parcel of land from them when they were looking for a place to move a house that was going to be demolished in downtown Amherst, Massachusetts. When my brother and I came along in the early 1980s, Peggy and Eddie encouraged us to treat the wetland as an extension of our much smaller backyard. I treasured this squelchy patch of land with its soft carpets of moss and vast expanses of mud and treated it as my playground. A shy kid, I found sanctuary in the wetland. The hours I spent barefoot in that patch of mud and bugs and rotten stumps, surrounded by the stench of aptly named skunk cabbage, using sticks to create elaborate floor plans for wetland architecture, provide one origin story for this book.

The academic roots of the project began in the English Department at Boston University. I can think of no better role model in the profession, nor a more generous and supportive advisor than Maurice Lee. Mo read countless drafts—always at startling speed—and has continued to give brilliant, helpful feedback in the intervening years. His support made the process of writing a dissertation more enjoyable and rewarding than I could have imagined, and I can only hope to pay it forward. Laura Korobkin turned me toward nineteenth-century women writers and has been a model of feminist scholarship. Hunt Howell was always willing to talk through inchoate ideas, and I owe many breakthroughs to our conversations. His friendship has been vital in the subsequent years. Anna Henchman and Joe Rezek prepared me for the profession and showed great kindness in the process. My deep thanks too to Julia Brown, Aaron Fogel, Susan Mizruchi, Tom Otten, and Anita Patterson and to fellow grad students who made coursework and dissertating so rewarding: John Levi Barnard, Heather Barrett, Iain Bernhoft, Andrew Christiansen, Paul Edwards, Christian Engley, Emily Donaldson Field, Reed Gochberg, Heather Holcombe, Emily Griffiths Jones, Claire Kervin, Niki Lefebvre, Liam Meyer, Caroline Oliver, Casey Riley, Emily Mohn Slate, and Arielle Zibrak.

For funding, I am grateful to the Boston University Center for the Humanities for a fellowship that let me conduct research in London and Paris. An ACLS/Mellon Dissertation Fellowship gave me the time and resources to complete a first version of this project. Archivists and librarians at Kew Library, the Royal Botanical Society, and Houghton Library steered me in fruitful directions. A San Gabriel Fellowship at the Huntington Library allowed me to explore the archives and the property's vast "living collection." And a Mellon SHASS Postdoctoral Fellowship at MIT granted me key time to reframe the project and develop new ideas. At UVA a Mellon Faculty Fellowship through the IHGC provided both time and a wonderful intellectual community for workshopping material.

A shortened version of chapter 3, "Dickinson and the Politics of Plant Sensibility," appeared in *ELH*, 85, no. 1 (Spring 2018): 41–70. A version of chapter 4 appeared in *American Literature* as "Garden Variety: Botany and Multiplicity in Harriet Beecher Stowe's Abolitionism," 87, no. 3 (September 2015): 489–516 and is reprinted here by permission of Duke University Press. At NYU Press, Priscilla Wald has been unflaggingly generous with her support at every turn. The book has greatly improved under her careful guidance and the support of series co-editors David Kazanjian and Elizabeth McHenry. The generous feedback from two anonymous readers made this a far better book. The staff at NYU Press have been incredibly helpful throughout the publication process. Thanks especially to my editor Eric Zinner for his support of the project, to Furqan Sayeed for responding to queries great and small, to Martin Coleman for guiding the production process, to Joseph Dahm for the careful copyediting, and to Mary Beth Jarrad for help marketing the finished product.

I feel grateful for all the places where I received feedback on the project over the years. Friends in Harvard's History and Literature concentration taught me much about the rewards of interdisciplinary work. Thanks especially to Lauren Kaminsky, Angela Allan, Nicole Eaton, Alex Orquiza, and Jen Schnepf. The year I spent as a postdoctoral fellow at

MIT shifted the foundations of the project and sent me back to the proverbial drawing board in the best possible way. Within the Literature section, Marah Gubar instantly made me feel at home and exemplified the power of postgraduate mentorship. I feel fortunate to call her and Kieran Setiya friends. Sandy Alexandre and Ruth Perry both sharpened my thinking as they made me feel like I belonged in 14N, and Ruth even entrusted me with the care of Big Boy. Harriet Ritvo treated me like an adopted member of the HASTS program. Mary Fuller, Wyn Kelley, Diana Henderson, Joaquín Terrones, Arthur Bahr, Shankar Raman, and James Buzard offered key support as well. Daria Johnson and Alicia Mackin helped me navigate a new institution. And I learned much from my fellow early career scholars who overlapped in the department: Joshua Bennett, Michaela Bronstein, and Rosa Martinez.

The First Book Institute was a transformative experience for workshopping the book, and I am so grateful to Sean X. Goudie and Priscilla Wald for creating this space and to CALS at Penn State for supporting it. Priscilla and Sean's generosity is hard to overstate, and they made the seemingly impossible possible through feedback, support, advice, and belief in the project's potential. I likewise learned so much from the other participants—Ben Bascom, Jordan Carroll, Christopher Perriera, Kathryn Walkiewicz, Sunny Xiang, Xine Yao, and especially Julianna Chow. I'm grateful too for Jeffrey Nealon's feedback. Participants at conferences where I presented parts of this project likewise helped shift the direction of my thinking. For feedback and conversations, I'm thankful to Sari Altschuler, James Finley, Lauren LaFauci, Sonia DiLoreto, Erica Hannickel, Ryan McWilliams, and Michelle Neely. My thanks too to Tim Hamilton and the Global Environmental Speaker Series at the University of Richmond. And I feel particularly indebted to the scholars who participated in the Living with Plants symposium in 2018 for several days of extraordinary interdisciplinary conversations: Courtney Fullilove, Theresa Kelley, Gabriela Soto Laveaga, Michael Marder, Annie Merrill, Cassandra Quave, and Tony Perry.

I finished this book at UVA in the company of incredibly supportive colleagues. My thanks especially to Anna Brickhouse for being such a wonderful mentor and helping me strategize a way forward on so many occasions. Elizabeth Fowler and Victor Luftig have sustained my family with trips to the magical Saddle Hollow. For feedback, advice, conversation, and myriad forms of support, my sincere thanks to Steve Arata, Alison Booth, Steph Ceraso, Karen Chase-Levenson, Sylvia Chong, Elizabeth Fowler, Susan Fraiman, Debjani Ganguly, Jennifer Greeson, Cristina Griffin, Bruce Holsinger, John O'Brien, Emily Ogden, Caroline Rody, Sandhya Shukla, and Brian Teare. Colette Dabney holds the department together and has been a source of friendship and wisdom throughout my time here. Heidi Siegrist was a phenomenal research assistant during my first years at UVA, and Henry Tschurr has helped me at the end of this process. I have also been fortunate at UVA to find community across disciplines. The Environmental Humanities working group and Environmental Thought and Practice advisory committees have been especially supportive groups, and my thanks especially to Willis Jenkins, Jim Igoe, Adrienne Ghaly, Enrico Cesaretti, Deborah Lawrence, Charlotte Rogers, Cassandra Fraser, Paul Freedman, and Dorothe Bach. Vivian Thomson welcomed me into the Environmental Thought and Practice major and has been a champion ever since. In the Architecture School, Beth Meyers, Andy Johnson, Erin Putalik, and Jessica Sewell have been particularly supportive. My thanks too to UVA's extraordinary librarians, especially Krystal Appiah, Sherri Brown, Christine Ruotolo, and Molly Schwartzburg. Before the pandemic, my writing group pulled me over writing walls and kept me accountable. Thanks in particular to Ira Bashkow for always taking notes and cheering us all on, and to Tess Farmer, David Singerman, Caitlin Wylie, Samhita Sunya, Fiona Greenland, Feyza Burak-Adli, and Jamie Jirout for all those poms.

A number of friends have shaped the ideas here either through conversation, meals, tea, walks, runs, and adventures: Rob Ambrose, Annika Swanson Berman, Steph Bernhard, Anouska Bhattacharyya, Jesse Billingham, Christina Black, Will Braff, Alex Brighton, Evan Bruno, Jenna Commito, Aly Corey, Maggie Dietrich, Maura Ewing, Ezra Feldman, Jennifer Hair, Greg Hamm, Austin Hetrick, Dom Higgins, Leonie Higgins, Adriana Paice Kent, Simon Kent, Georgiana Kuhlmann, Kyrill Kunakhovich, Hana Lewis, Anne-Garland Mahler, Laura Martin, Kara Mazzotta, Gwendolyn McDay, Marie McDonough, Lisa Messeri, Jyothi Natarajan, S. Aykut Ozturk, Heidi Hayes Pare, Claire Payton, Maya Ray, Matt Reardon, Chris Rhie, Kathryn Schwartz, Vanessa Scurci, Chet'la Sebree, Nora Segar, Dan St. Jean, Dijit Taylor, Emily Taylor, and Anand Vaidya.

In London Meriel and Jeremy Greenhalgh graciously housed me for repeated trips to the archives, and I wrote a good deal under their eaves. Kamilla Arku has been a dear friend across two decades and continents. Meeting Feng-Mei Heberer at MIT was a stroke of good luck. Casey Riley has always helped me see the bigger picture. Emily Mohn Slate's kindness and intuition are nonpareil. Ever since our first encounter with a skunk, Sarah Milov has made life in Charlottesville more exciting. Sara Torres has helped me sift through and sharpen muddled ideas and championed progress. Caterina Scaramelli and Ben Siegel have hosted, fed, and kept me company across years and miles; I'm grateful for their friendship and our Monday pizza dinners. Anna Berman has talked through every chapter of the book, often over ice cream or while hiking. The book is far better for her friendship, as am I. Benny Taylor and Cricket Arrison gamely agreed to hang out with me in the belly of the whale with their A+ cetology humor. The first inklings for this book were born out of long conversations in graduate school with Arielle Zibrak, and to riff off Frank Bidart, I cannot conceive of its face without her.

Finally, the book would not exist without my family's love and steadfast support. I wish my grandmothers Mary and Polly were still around to see it. Noah rescued documents more times than I can count, and he, Kate, Anna, and Brooks have ensured the dance mix was always queued up. My thanks too to Mindy, Paul, Kevin, my late aunt Debbie, my late grandpa Jack, Megan, Brian, Hannah, Becca, and Gabe, and to Adam, Kathy, Phil, Jeff, Martha, Bryon, Helen, Jim, Joe, Lily, and Nate. Kathy Roth and Phillip Singerman have been incredibly supportive, celebrating progress and ensuring precious weekend hours to work. My parents, Jeanie and John Kuhn, let me play out in the mud far after dark and took me into the mountains before I could walk. Their home has always been full of love, and their support has been unwavering over the years. Any gratitude expressed to them here is simply insufficient.

I met David Singerman as we were both finishing our dissertations. His kindness, wit, generosity, humor, patience, and love have brightened every day since. David has read every word, accompanied me on research trips and running paths, and provided an endless stream of baked goods. He inspires and grounds me. This book is for him, and for our daughter Marguerite. Sharing the everyday with them is my greatest good fortune.

NOTES

INTRODUCTION

1. Phelps was widowed twice. At the time she first published *Familiar Lectures* she did so using her late first husband's last name, Lincoln. When she remarried in 1831 she took her second husband's last name, Phelps. Critical consensus coheres around referring to her by this latter surname, and so to avoid confusion I refer to her as Phelps throughout.

2. Baym, *American Women of Letters*, 18.

3. One in the *New York Mirror* described how the book conveyed "the dignity, the value, and the auspicious tendencies upon morals, of the science of botany" and reflected "credit on our common country . . . well calculated to enhance its estimation among scientific men in Europe." *New York Mirror*, reprinted in the *Ladies' Literary Portfolio* 1, no. 50 (November 25, 1829).

4. Phelps, *Familiar Lectures on Botany*, 12.

5. Translation is Andrea Wulf's. See *Invention of Nature*, 88.

6. Ghazal Jafari, Pierre Bélanger, and Pablo Escudero have written about the extremely destructive effects of the map Humboldt produced in 1805, *Geographie des Plants Équinoxiales*, facilitating the extraction of Cinchona and territorial dispossession. See *A Botany of Violence*.

7. Chakrabarty, *Provincializing Europe*; Chakrabarty, "Climate of History"; De-Loughrey, *Allegories of the Anthropocene*. On the profound environmentalist legacy of this, see Guha and Martinez-Alier, *Varieties of Environmentalism*. On colonial gardens, see Casid, *Sowing Empire*; Bleichmar, *Visible Empire*. See also Batsaki, Cahalan, and Tchikine, *Botany of Empire*.

8. See Allewaert's *Ariel's Ecology*, which powerfully illuminates how resistance to colonial control could be forged in an ecological embrace of tropical climates and topographies. See also Karen Kupperman's "Fear of Hot Climates." Tony Perry has also described how enslaved people mobilized environmental conditions against the plantation system. And in *Undisciplined* Nihad Farooq has shown how definitions of subjectivity and personhood fluctuated in the context of nineteenth-century race science.

9. For a discussion of this see, for example, Tony Perry's "In Bondage When Cold Was King." See also Andreas Malm's "In Wildness Is the Liberation of the World."

10. Allewaert describes how the "closely related discourse of sympathy and interest depend on the fantasy of the bounded body that may, through the working of

sympathy, be motivated to care for another." See *Ariel's Ecology*, 18. On sentiment and racial construction, see Schuller, *Biopolitics of Feeling*.

11. Amy Kaplan noted in *Anarchy of Empire* that domestication, "conquering and taming the wild, the natural, and the alien" (25), was a key function of domesticity.

12. On seeds as political objects, see Fullilove, *Profit of the Earth*. The relationship between domestic gardening and political power in the U.S. national context has been much harder to see than in the imperial context (see, for instance, Bewell, *Natures in Translation*; Kelley, *Clandestine Marriage*; Tobin, *Colonizing Nature*; Casid, *Sowing Empire*; Schiebinger, *Plants and Empire*).

13. For example, Harriet Beecher Stowe kept such a glass case in her bedroom. Sarah Mapps Douglass painted intricate drawings of flowers in friendship albums. Sophia Hawthorne kept a plant drawer to instruct her children.

14. Of course, a number of figures most associated with romantic wilderness formulated wildness at times as a kind of home. Thoreau and Muir are both good examples of this. See D'Amore, "Thoreau's Unreal Estate." See also Walls, *Henry David Thoreau*.

15. Several scholars have shown how planters anxiously responded to environmental uncertainty. Lynne Feely describes how the unruliness of plants threatened planters' authority, tracing a phenomenology in which "planters could not cede authority to the land lest they concede that enslaved people also escaped total control, an idea that hit on the region's deep anxiety that its slaves would rise up against it" ("Plants and the Problem of Authority," 54). Susan Scott Parrish describes how planters mediated the environmental risk to their own selves, property, and wealth by pushing this risk onto their enslaved labor force. As Parrish puts it, plantation powers invented race "to make a buffer zone between itself and . . . biological uncertainty. . . . Whites protected themselves, their property, and the category of whiteness itself by positioning black bodies to absorb various manifestations of biological chance: pest infestations, unpredictable weather, infertile soil, tropical earthquakes, fluctuating crop yields, contagious disease, unexpected fires in sugar manufacture, and risky oceanic transit." See *Flood Year 1927*, 8–9.

16. Mark Rifkin has described how the settler state inscribes its power so as to seem ordinary or commonsensical. See *Settler Common Sense*.

17. Philip Pauly makes this point in his landmark study of horticulture in the nineteenth-century United States. See *Fruits and Plains*. He urges that "we need to set aside the century or more of discussions about this keyword among literary critics and anthropologists to recover a world where culture was an umbrella term for efforts at biotic improvement. Horticulturalists could speak unself-consciously about strawberry culture and pear culture. For them, high culture meant neither Plato nor fancy table settings, but rather well-rotted manure and hand weeding. Yet they understood that their usages shaded smoothly into others: that pear culturalists were themselves a culture; that geography, law, and race were, with climate and seeds, elements that made cotton culture possible; that culture happened not only on farms, but in schools, churches, and their own clubs" (6). Liter-

ary writers produced a similar slippage and set of connections in their writing, as I will show.

18. As Susan Fraiman reminds us, we should pay as much attention to the lived realities of domestic life as to the domestic ideologies that prescribed behaviors and values. See *Extreme Domesticity*.

19. Lisa Lowe notes that "the human" itself is an exclusionary product of liberalism: "The social inequalities of our time are a legacy of these processes through which 'the human' is 'freed' by liberal forms, while other subjects, practices, and geographies are placed at a distance from 'the human.'" See *Intimacies of Four Continents*, 3. See also Weheliye, *Habeas Viscus*. As Jeffrey Nealon has noted, plants have played an essential, if ignored, role in Western metaphysics; they are integral to the construction of biopower. Nealon, *Plant Theory*.

20. This moves away from Romantic holism, allowing for different ways of seeing than that governed by the relationship between part and whole. For an excellent perspective on partiality and difference as loopholes in settler colonial geographies, see Juliana Chow's *19th Century American Literature*. DeLoughrey's *Allegories of the Anthropocene* also powerfully illustrates the importance of thinking about particular localities. And Ursula Heise's *Sense of Place* uses the term "eco-cosmopolitanism" to address interconnection across different ecological spaces and scales.

21. Technologies for sustaining plant life became mediums for imagining a global environment. In their best-selling *The American Woman's Home* (1869), Harriet Beecher Stowe and her sister Catharine Beecher instructed readers on how to construct a simple Wardian case and fill it with material found "by searching carefully the rocks and clefts and recesses of the forest." But they later quoted Dr. Oliver Wendell Holmes's description of the terrestrial world as a Wardian case writ large: "A great mist of gases and of vapor rises day and night from the whole realm of living nature. . . . The salt-water ocean is a great aquarium. The air ocean in which we live is a 'Wardian case,' of larger dimensions." Holmes's description of the earthly world as an "air ocean" and "Wardian case" conjures an image of an environment understood through the technology of a glass case, one that was rapidly altering biotic transfer around the globe. In the context of the Beecher sisters' domestic manual, this image makes the globe a projection of the domestic interior composition.

22. Courtney Fullilove has described how the United States' expanded commerce around the globe in the 1850s gave rise to various kinds of global imaginations: "as a patchwork of militarized nation-states, a hierarchy of civilizations, a grid of marketplaces, and a zone of common nature" each requiring different "rules of conduct." See *Profit of the Earth*, 28.

23. Kaplan, *Anarchy of Empire*, 13.

24. Ward, *On the Growth of Plants*, 23.

25. Animals were also bred in the colonies for imperial consumption. Rebecca Woods's *Herds Shot Round the World* illustrates how colonial livestock fed the British metropole.

26. See Braun, "Bioprospecting Breadfruit."
27. On colonial gardens, see Casid's *Sowing Empire* and Bleichmar's *Visible Empire*.
28. James Scott has described how domestication was fundamental to state power and how it required a subordinated population. Ruth Wilson Gilmore has described how in the context of empire and the rise of capitalism racism became a way to justify inequality: "Capitalism requires inequality," she notes, "and racism enshrines it" (Card, *Geographies of Racial Capitalism*).
29. Ward, *On the Growth of Plants*, ix.
30. On the nineteenth-century "annihilation of space and time," see Rebecca Solnit, *River of Shadows*.
31. Romanticists, environmental humanists, and postcolonial scholars have all identified the imperial circulation of plants starting in the early modern period as an "early phase of globalization" (DeLoughrey, *Allegories of the Anthropocene*, 24). New technologies for traversing the globe at speed and for making nature "portable" not only rapidly shifted the material realities across continents, but also reshaped how individuals conceptualized their own situated knowledges in relation to other parts of the world.

    In describing the history of breadfruit in the Americas, Juliane Braun describes how colonized spaces "that seem disconnected and far apart not only influenced each other but also changed imperial policies in the metropole." "Bioprospecting Breadfruit," 646. See also Bewell, *Natures in Translation*. See the introduction to Elizabeth DeLoughrey's *Allegories of the Anthropocene* for a more detailed genealogy. DeLoughrey credits Mary Louise Pratt with emphasizing this point in *Imperial Eyes*.
32. Emily Dickinson to Dr. and Mrs. J. G. Holland, Letter 315 (Early March 1866), in *Selected Letters*, 191.
33. In the British context Alan Bewell notes how portable nature in the form of natural history specimens and objects helped shore up an essentialist idea of "English" nature. See "Romanticism and Colonial Natural History," 13.
34. Describing the diversity of natures across the globe, Bewell emphasizes that "every nature . . . is an expression of the politics of the people who depend on it" (*Natures in Translation*, 15).
35. Jefferson, *Notes on the State of Virginia*, query IV, 61.
36. See chap. 1 in Pauly, *Fruits and Plains*.
37. Thoreau, *"Wild Apples" and Other Natural History Essays*, 71.
38. The point that U.S. empire temporally aligns with "the birth of the United States" is foundational to indigenous studies, as Jodi Byrd notes, and should be foundational to American studies. See Byrd, *Transit of Empire*, xiii.
39. See Pauly, *Fruits and Plains*; see also Wulf, *Founding Gardeners*.
40. Cultivation, to paraphrase Sonya Posmentier, was catastrophic. See Posmentier, *Cultivation and Catastrophe*.
41. As scholars like Monique Allewaert, Judith Carney, Christopher Iannini, Susan Scott Parish, and Londa Schiebinger have shown, bioprospectors and colonists

alike had historically relied upon Indigenous and enslaved people's knowledge of plants, particularly when it came to healing diseases against which European pharmacopeia was ineffective. Sharla Fett illustrates the expertise that enslaved workers had with "working cures" that combined botanical expertise with religious traditions. See Allewaert, *Ariel's Ecology*; Carney, *Black Rice*; Carney and Rosomoff, *In the Shadow of Slavery*; Iannini, *Fatal Revolutions*; Parish, *American Curiosity*; Schiebinger, *Plants and Empire*; Fett, *Working Cures*.

42. Courtney Fullilove has characterized this process as similar to the way that biodiversity collection missions today often prioritize gathering plants' genetic sequencing over forms of local knowledge. See *Profit of the Earth*.

43. *American Agriculturalist*, "Visit to the Royal Gardens at Kew," 518. The author also praised the mission of the gardens, "open to the well-behaving public" and, quoting John Lindley, "designed chiefly to promote the advancement of science and the arts, of medicine, commerce, agriculture, horticulture, and various branches of manufactures."

44. Fullilove, *Profit of the Earth*, chaps. 1–2.

45. In *Anarchy of Empire* Amy Kaplan argued that the term "domestic" "has a double meaning that links the space of the familial household to that of the nation, by imagining both in opposition to everything outside the geographic and conceptual border of the home" (25). Scholars of nineteenth-century domesticity in both the United States and England have illustrated the complexities of domestic relations and domestic politics. See Merish, *Sentimental Materialism*; Romero, *Home Fronts*; Armstrong, *Desire and Domestic Fiction*; Freedgood, *Ideas in Things*. Susan Fraiman's recent *Extreme Domesticity* productively decouples the lived experiences of domesticity from domestic ideology.

46. Phelps, *Familiar Lectures on Botany*, 12.

47. Amy King describes the way that Victorian novelists correlated "blooming" young women with a sexualized courtship plot. See *Bloom*. Deirdre Lynch has illustrated how this trope accommodated new developments in greenhouse technology that shifted the temporality of blooming. As organic time was shaped by new technologies that allowed cultivators to "force" plants or grow them year round in temperature controlled environments, the "natural development" of female protagonists within courtship plots might become "idiom[s] of artifice" (719). Lynch, "'Young Ladies Are Delicate Plants.'"

48. See Gianquitto, *"Good Observers of Nature."*

49. Phelps, *Familiar Lectures on Botany*, 11.

50. In addition to Gianquitto, Nina Baym and Elizabeth Keeney have shown the substantial cultural impact of these lessons, particularly for women. See Baym, *American Women of Letters*, and Keeney, *Botanizers*.

51. Bartram, *Travels*, li.

52. Phelps, *Botany for Beginners*, 9.

53. Phelps, *Botany for Beginners*, 9.

54. Phelps, *Botany for Beginners*, 12.

55. *Youth's Companion*, "Variety: Poisonous Plant."

56. Woodlin, "Communication."

57. *American Agriculturalist*, August 1, 1847, 261; *Godey's Lady's Book*, July 1, 1847, 57.

58. Stowe, "Introduction."

59. On portability, see Bewell, "Romanticism and Colonial Natural History," 17.

60. Ritvo, "At the Edge of the Garden," 364.

61. Thoreau, *Writings of Henry David Thoreau*, 219, 343, 219.

62. *Putnam's Monthly Magazine*, "Chat about Plants," 428.

63. Thoreau, *"Wild Apples" and Other Natural History Essays*, 59. Of course, "Walking" has manifest destiny sympathies, and in this light "being at home everywhere" could imply an imperial fantasy.

64. Cosmopolitanism is a fraught term, especially considered through the light of European intellectual and imperial history. Here I draw on the context theorized by Breckenridge, Pollock, Bhabha, and Chakrabarty in their introduction to the edited collection *Cosmopolitanism*: "ways of living at home abroad or abroad at home—ways of inhabiting multiple places at once, of being different beings simultaneously, of seeing the larger picture stereoscopically with the smaller" (11).

65. An article in *Scribner's Monthly* called "Curiosities of Plant Life" describes how "narcotics paralyze and poisons kill [plants]. Electricity stimulates or stuns them. . . . De Candolle placed lightly a drop of water on a leaflet of a sensitive plant. No motion followed. He touched it with a drop of acid, and on the instant the leaflets shrunk and drooped" (651).

66. See Browne, *Charles Darwin*.

67. See Armbruster and Wallace, *Beyond Nature Writing*; Ruffin, *Black on Earth*; Smith, *African American Environmental Thought*; Kilcup, *Fallen Forests*; Glave and Stoll, *"To Love the Wind and the Rain."*

68. Hawthorne, *Passages from the American Notebooks*, 97.

69. As Alan Bewell has recently argued, "There is no Nature—only natures—and . . . most of these have undergone translation in time or across space." The upshot to reframing in this way is "a major shift in how the natural world was understood, from a belief in a nature that was histories shaped by mobility, conflict, and change" (*Natures in Translation*, xiv). This tension between particularity and universality has reappeared in current Anthropocene debates. See Chakrabarty, "Climate of History"; DeLoughrey, *Allegories of the Anthropocene*; Yusoff, *Billion Black Anthropocenes*; Whyte, "Indigenous Climate Change Studies"; Baucom, *History 4 Celsius*; and LeMenager, "Love and Theft."

70. Wampole, *Rootedness*.

71. This is despite, as he notes, a "conceptual allergy" to plant life in Western philosophy (see Marder, *Philosopher's Plant*).

72. Animal studies is a well-established field at this point. For some early influential studies, see Ritvo, *Animal Estate*; Ritvo, *Noble Cows and Hybrid Zebras*; and Haraway, *When Species Meet*. For an example of how literary scholarship has recently shaped the field, see Bennett, *Being Property Once Myself*.

73. See Laist, *Plants and Literature*. Catriona Sandilands describes this power of plants in her review essay "Fear of a Queer Plant." While I believe that queer theory can be helpful in thinking about the role of plants in the texts that follow, I largely avoid using the term in order to focus on the vocabulary that nineteenth-century actors used to describe their relationship with plant life. In this I follow Greta LaFleur's excellent model in *The Natural History of Sexuality* in thinking through the "precipitousness" of contemporary analytics and vocabularies for past experiences.

74. Casid, *Sowing Empire*.

75. As Casid notes, "Plant crossbreeding, transfer, and reclimatization were at the center of British efforts to control world markets in highly valuable vegetable commodities" (*Sowing Empire*, 3).

76. See Martin, Myers, and Viseu, "Politics of Care in Technoscience," 626.

77. In her influential study unpacking the biopolitics of feeling in the nineteenth century, Kyla Schuller has emphasized how "accounts claiming that the liberatory potential of mobility and plasticity can redress what they see as the fixed taxonomies of nineteenth-century racial personhood misrecognize one of biopolitics' most pervasive effects. Biopower works by situating individuals in dynamic relation and calculating and regulating how their bodies affect one another within a milieu" (*Biopolitics of Feeling*, 11).

78. Of course these are not issues limited to the Eastern Seaboard. The "discovery" of valuable plants in the western part of the continent during scientific explorations such as the Lewis and Clark Expedition justified expansionist efforts. In Maria Ruiz de Burton's 1885 *The Squatter and the Don* fruit trees are commercial crops with significant land use implications. And as Erica Hannickel has argued, botanists like George Engelmann in the Midwest were deeply imbricated in transnational networks of exchange. Mauro Ambrosoli has likewise traced the introduction of alfalfa into the United States from two directions: it entered the East Coast as luzerne and the West Coast as Chilean alfalfa (introduced in the Andes by the Spanish in the sixteenth century). See Ambrosoli, "From Europe to the Americas."

79. Dickinson, *Poems of Emily Dickinson*, F256 A.

1. BOTANICAL NATIONALISM

1. Ceylon became the republic of Sri Lanka in 1972. Winterthur curators have dated the cup to roughly sometime between 1800 and 1840, a period during which Britain colonized Ceylon. For the museum's entry on the cup, see Object Number 1965.1836 A, B, "Cup (Covered Cup)," http://museumcollection.winterthur.org.

2. For a discussion of "it" narratives, see the edited collection by Blackwell, *Secret Life of Things*.

3. See museum blog post by curatorial fellow Lea Lane, "Lovely Bunch of Coconuts."

4. See Schiebinger, *Plants and Empire*, 5.

5. Other important work on this topic includes Endersby, *Imperial Nature*; Crosby, *Ecological Imperialism*; Cronon, *Changes in the Land*.

6. In the late 1860s one coconut cup in particular made the news in agricultural periodicals in the United States. The carved cup, belonging to Alexander Selkirk, upon whom Defoe based Robinson Crusoe, was donated to the Antiquarian Society of Scotland. See "Robinson Crusoe," *Scotsman*, reprinted in *New England Farmer, and Horticultural Register*, August 14, 1869, 48, 33, 4.

7. For an excellent discussion of women in the sciences in nineteenth-century America, see Nina Baym, *American Women of Letters*. For a discussion more tailored to botany, horticulture, and plant studies, see Gianquitto, *"Good Observers of Nature"*; Keeney, *Botanizers*. And Beverly Seaton's *Language of Flowers* provides the best introduction to the way in which the culture surrounding language of flower books in America, England, and France helped to shape and participate in sentimental discourse.

8. For a comprehensive biography of Child's life and work, see Karcher, *First Woman in the Republic*.

9. Child's first novel, *Hobomok*, published the same year as *Evenings in New England*, uses a frame narrative to articulate the need for distinctively American literature. Her popular 1829 household manual *The American Frugal Housewife* appended the word "American" to distinguish it from the British manual *The Frugal Housewife*.

10. Phelps, *Familiar Lectures on Botany* (1854), 5, 10.

11. Beecher, *Treatise on Domestic Economy*, 56.

12. Phelps, *Familiar Lectures on Botany* (1854), 18.

13. Child, *Evenings in New England*, 25. Subsequent references to this text appear parenthetically and refer to this edition.

14. Jim Endersby discusses the extent to which Joseph Hooker's botanical collection process relied upon an extensive network of botanical collectors: people in the field who did the actual work of collecting. This involved a far greater cast of characters than credited in these kinds of prefatory remarks, and interestingly, as Endersby notes, "as some of these collectors worked to improve their expertise, they came to realize that it was their metropolitan correspondents who were becoming dependent—since accomplished collectors were hard to find." See *Imperial Nature*, 55. The director's correspondence at Kew is full of records of this kind that provide insight into collection practices. Take, for instance, a letter from James Waters to Joseph Hooker on October 6, 1851: "Dear Sir William," he begins, "Having received a supply of seeds of some of very beautiful flowering plants from Jamaica by my daughter-in-law, who is on her way to Bruges, I take the liberty of sending some of the most rare by her, thinking you may not be in possession of them." Waters, "Letter from James Waters to Sir Joseph Hooker."

15. Lewis, "Gathering for the Republic," 69.

16. Lewis, "Gathering for the Republic," 71.

17. Crevecoeur, *Letters from an American Farmer*, 46, 14.

18. Of course, when nineteenth-century botanists sought to identify the origin of particular species and varieties of plants in North America, their assignations

sometimes naturalized plants that had been brought over to the continent in 1492. As Alfred Crosby has described, the Columbian Exchange shaped the global distribution of plant matter. This was partially the result of strategic human effort, as in the way that Europeans sought to possess, cultivate, and capitalize upon economically useful plants. See Crosby, *Columbian Exchange*. Contact also led to unanticipated biotic exchanges. Earthworms, for instance, arrived in North America when the Virginia colonists started selling tobacco to England. As British ships unloaded ballast of English soil to make room for barrels of tobacco, they unknowingly unleashed worms that transformed the local ecology. See Mann, *1491*, 40.

19. Nuttall, *Genera of North American Plants*, vi.
20. For more information on publication history, distribution, and popularity, see Baym, *American Women of Letters*, 18; see also Gianquitto, *"Good Observers of Nature,"* 22.
21. Phelps, *Familiar Lectures on Botany* (1854), 3, emphasis original.
22. For more on Romantic science in a transnational context, see Holmes, *Age of Wonder*.
23. Lydia Howard Sigourney, *Letters to Young Ladies*, 34.
24. Sigourney, *Letters to Young Ladies*, 34, 35.
25. Lambert, *American Forest*, 8.
26. Lambert, *American Forest*, 9.
27. Lambert, *American Forest*, 144.
28. Likewise, articles on rubber were featured in periodicals like *Every Youth's Gazette, The American Magazine of Useful and Entertaining Knowledge, Massachusetts Ploughman and New England Journal of Agriculture, The Anglo American,* and *Horticulturalist and Journal of Rural Art and Rural Taste*, among many others.
29. *Robert Merry's Museum*, "Wonderful Trees, No. 3," 131, 132.
30. Dearborn, "Address Delivered before the Massachusetts Horticultural Society," 18.
31. Child, *Evenings in New England*, 23.
32. For a good discussion of this, see Tamarkin's *Anglophilia*.
33. Kilcup, "Feeling American in the Poetic Republic," 303.
34. Sigourney, *Voice of Flowers*, 94.
35. Sigourney, *Voice of Flowers*, 95.
36. Sigourney, *Voice of Flowers*, 96.
37. Hale, *Flora's Interpreter*, iii.
38. Loughran, *Republic in Print*, 3.
39. Hale, *Flora's Interpreter*, iii.
40. Hale, *Flora's Interpreter*, 46.
41. Hale, *Flora's Interpreter*, 56.
42. As Sari Altschuler has argued, physicians' own narratives have tended to dominate the history of medicine in the late nineteenth century, largely distorting our historical view as to the significance of herbal remedies and other alternatives to professional—white male—practice. See Altschuler, *Medical Imagination*.

43. As Karen Kupperman notes, "Early modern science taught that human beings and their native physical environment normally existed in a state of ecological harmony" (213). Kupperman, "Fear of Hot Climates."

44. Kupperman, "Fear of Hot Climates," 225.

45. Valencius, "The Health of the Country," 169.

46. Valencius, "The Health of the Country," 179.

47. Cooper, Experienced Botanist, iv, v.

48. Cooper, Experienced Botanist, v–vi.

49. Cobbett, American Gardener, 8, emphasis original.

50. Cobbett, American Gardener, 8.

51. Downing, Treatise on the Theory and Practice of Landscape Gardening, 1.

52. Downing, Treatise on the Theory and Practice of Landscape Gardening, 2, emphasis original.

53. Downing, Treatise on the Theory and Practice of Landscape Gardening, 3.

54. Downing, Treatise on the Theory and Practice of Landscape Gardening, 4.

55. Amy King describes how Linnaeus's sexual classification of plants in the mid-eighteenth century became the basis for specifically discussing a young girl's sexual maturation, courtship, and marriage. She calls this the "bloom" plot and locates it in Victorian texts across the nineteenth century. See Bloom.

56. The floral trope as an analogy for human affairs has a long history. For instance, Shakespeare's The Winter's Tale (1623) contains a significant discussion of grafting and interbreeding. In act 4, scene 4, Polixenes replies to Perdita, "we marry / A gentler sien to the wildest Stocke, / And make conceyue a barke of baser kinde / By bud of Nobler race" (109–112). Erasmus Darwin's "The Loves of Plants" (1789) is an influential example of a scientific text that analogized humans and plants, particularly along reproductive lines. See Darwin, Botanic Garden. Theresa M. Kelley details the way in which various methods of botanical classification influenced social organization during the Romantic period in Britain. See Clandestine Marriage. The comparison between masculinity and flowers has been explored less, though for a discussion of masculinity, race, and sexuality, see Looby, "Flowers of Manhood."

57. Describing Victorian culture, Amy King notes the centrality of "the girl 'in bloom,' or the female whose social and sexual maturation is expressed, rhetorically managed, and even forecast by the use of a word (bloom) whose genealogy can be traced back to the function of the bloom, or flower, in Linnaeus's botanical system" (Bloom, 5).

58. Many critics have rightly noted that despite—and in part because of—aggressive anti-Indian policies and rampant racism, the figure of the Indian played an important role in white efforts to establish national distinctiveness. For detailed treatments of this topic, see Slotkin, Regeneration through Violence; Trachtenberg, Shades of Hiawatha. Sherry Sullivan notes that Native Americans in early fiction had a "symbolic function . . . [that was] . . . twofold: not only as an anti-image against which Americans distinguished themselves, but also as a positive image

with which they sought to be associated." See Sullivan, "Redder Shade of Pale," 57. Linda Kerber likewise stresses the point when she writes, "The Indian issue . . . was a popular way . . . by which fiction writers responded to urgent demands for the creation of a national literature; what more authentic emblem of the American experience could there be than the native American himself?" See Kerber, "Abolitionist Perception of the Indian," 271. See also chapter 1 of Mark Rifkin's *Settler Common Sense* for a discussion of works that have analyzed "U.S. imperial absorption of Native peoples and lands in nineteenth-century writing" primarily through a study of the way this literature represents Native figures.

59. Child, *Hobomok and Other Writings on Indians*, 8. Subsequent references to this text appear parenthetically and refer to this edition.

60. *New England Farmer, and Horticultural Register*, "Grafting Fruit Trees," 262.

61. Crevecoeur, *Letters from an American Farmer*, 205. Philip Pauly describes how Thomas Jefferson, at twenty-four, "recorded a patriotically meaningful if biologically wrongheaded experiment to improve the taste of native cherries by grafting wild American buds onto cultured Old World rootstocks." See *Fruits and Plains*, 2.

62. Paula Kot points to the fact that "in the relationship between Mrs. Mary Conant and her daughter, Child literalizes the metaphor of the body politic by continually erasing the boundaries between mother and daughter . . . their destinies are intricately intertwined, like the historical and aesthetic destinies of the Unites States and England." See Kot, "Engendering Identity," 93.

63. Wald, "Terms of Assimilation," 69. Speaking to the way that the novel constructs shared race as a means to obviate religious differences in the community, Ezra Tawil has likewise pointed out, "By marrying out and then returning to her birth community, Mary Conant changed what it meant to belong to that group in the first place. She left a community in which the difference between Christian sects was paramount. But the community to which she returned was one in which Episcopalians and Puritans shared something more fundamental: their race." See *Making of Racial Sentiment*, 113.

64. Laura Mielke argues, for instance, that "Child naturalizes racial segregation, and by locating Native Americans' future solely within Euro-American families and homes, she insists on erasing Native American culture and sovereignty." Hobomok's disappearance in this light is linked to "the ideology of removal: geographic segregation in accord with a racial-cultural hierarchy, overseen by benevolent Euro-Americans." See "Sentiment and Space," 173. Nancy Sweet writes along similar lines that Hobomok "represents the wild and untamable in America, and Child banishes him to make way for the erudite, enlightened, and European-like society that Mary and Charles Brown together establish in the New World." See "Dissent and the Daughter," 121.

65. See Mielke, *Moving Encounters*; Schuller, *Biopolitics of Feeling*; Tawil, *Making of Racial Sentiment*.

66. Tawil, *Making of Racial Sentiment*, 2, 11.

67. Byrd, *Transit of Empire*, xix.

68. See Slotkin, *Regeneration through Violence*; Trachtenberg, *Shades of Hiawatha*.

69. As Carolyn Karcher notes, despite the novel's participation in imperialist logic, it stands out for its exploration of "a radical alternative to race war and white male supremacy" (*First Woman in the Republic*, 34).

70. For a more in-depth discussion of *The First Settlers* and its politics, see Karcher, *First Woman in the Republic*, 87–90.

71. Samuels, "Women, Blood, and Contract," 60.

72. See Kades, "History and Interpretation."

73. Many of the best early works of environmental criticism have been critiques of the kinds of organizing ecological metaphors that I am arguing Child is actively working to construct. Perry Miller's *Errand into the Wilderness* (1956) explores wilderness as a figure of speech and addresses the significant way in which this conception of North America played a pivotal role in Puritan theology. Likewise, Leo Marx's technological pastoral in *The Machine in the Garden* (1964) illustrates the important ways in which the idea of nature grounded an exceptionalist ideology. The first line of that famous work asserts, "The pastoral ideal has been used to define the meaning of America ever since the age of discovery, and it has not yet lost its hold upon the native imagination." One could modify the terms here slightly and come close to Annette Kolodny's thesis in *The Lay of the Land* (1975): the idea of a feminized landscape has likewise been used to define the meaning of America. These works share the important premise that, historically, the idea of nature has been part of the project of national distinction. To this collection we might add a host of other critical works—Henry Nash Smith's *Virgin Land* (1970), Richard Slotkin's *The Fatal Environment* (1998)—attentive to the idea of the landscape as a vital organizing metaphor for American identity.

74. Karcher, *First Woman in the Republic*, 510.

75. Karcher argues that interracial marriage is a narrative solution that Child turns to throughout her career as a potential form of racial reconciliation. As she astutely notes, the inequality built into the institution of marriage limits the extent to which it could serve as an effective model for racial equality. See Karcher, *First Woman in the Republic*, 511. Cassandra Jackson argues that the novel constructs the interracial heroines as model "new Americans" who survive by performing for white patriarchy (49). See *Barriers between Us*. Dana Nelson focuses on the novel's critique of patriarchal culture, and especially its commodifying gaze. See Nelson, *Word in Black and White*.

76. Child, *Romance of the Republic*, 286. Subsequent references to this text appear parenthetically and refer to this edition.

77. Rosenthal, "Floral Counterdiscourse," 222.

78. Freedgood, *Ideas in Things*, 1.

79. Davidson, *Revolution and the Word*; Anderson, *Imagined Communities*; Warner, *Letters of the Republic*. For print culture and provincialism, see Rezek, *London and the Making of Provincial Literature*. On nationalism and early African American

print culture, see Spires, *Practice of Citizenship*; Fagan, *Black Newspaper and the Chosen Nation*.
80. Loughran, *Republic in Print*, 9.
81. Perry Miller popularized the term "nature's nation" to describe America. For an environmental discussion of this idea, see Daegan Miller, "Reading Trees in Nature's Nation." As Conevery Bolton Valencius writes, European and American migrants who moved west "saw themselves as participating in a process that was essentially *American*: taking over 'new' lands and remaking them into American territory," what she calls "a national as well as a personal project" (*"The Health of the Country,"* 10).

2. BOTANICAL DISRUPTION

1. *New England Farmer*, "Concord Grape," 161.
2. Pauly, *Fruits and Plains*, 74–76.
3. For a fuller account of Bull's creation of the Concord, see Pauly, *Fruits and Plains*.
4. Sophia Peabody Hawthorne, quoted in Valenti, "Sophia Peabody Hawthorne's *American Notebooks*," 141.
5. Pauly describes the ways in which the term "culture" in the nineteenth century had a strong link to actual humus (6). In the Concord context, Lesley Ginsberg identifies the way that reformers like Bronson Alcott used plants as a metaphor for socialization. Laura Dassow Walls has described the ways in which organicism (in which parts are subservient and instrumental to the whole) "offers growth without change, an evolutionary model that safely contains the revolutionary threat." See Walls, *Seeing New Worlds*, 76.
6. James, *Hawthorne*, 3, 5.
7. Michael Colacurcio outlines the perpetuation of this perception in the critical tradition, dividing Hawthorne critics into "Jameseans" and "Melvillians," both of whom situate Hawthorne as deeply engaged with a fundamentally local history and politics. See Colacurcio, *Province of Piety*. See also Milder, "Hawthorne and the Problem of New England."
8. James, *Hawthorne*, 117.
9. The Hawthorne that emerges once we contextualize his botanical engagement is far more cosmopolitan, more global in his thinking, than is usually understood. A recent string of critics such as Anna Brickhouse, Laura Doyle, and Robert S. Levine have responded to the Jamesian view by uncovering how Hawthorne's literary preoccupations place him within a transnational and hemispheric antebellum context. See Brickhouse, *Transamerican Literary Relations*; Doyle, *Freedom's Empire*; Levine, *Dislocating Race and Nation*. Andrew Loman describes how his work on monetary systems uncovers "a cosmopolitan Hawthorne." See Loman, "'More Than a Three-Pence,'" 357; see also Luedke, *Nathaniel Hawthorne and the Romance of the Orient*.
10. As Brenda Wineapple recounts, "In 1838 Manning published *The Book of Fruits: Being a Descriptive Catalogue of the Most Valuable of the Pear, Apple, Peach, Plum*

& Cherry, for New England Culture [and] in it . . . praises the Hawthorndean apple as a medium-sized fruit, remarkably handsome, flesh white and very juicy but not highly flavored: perhaps his perspective on Nathaniel. For try as he might, Robert Manning could not cultivate his nephew as he might an apple or peach." See Wineapple, *Hawthorne: A Life*, 37.

11. Dearborn, "New Fruits and Ornamental Plants," 324.

12. *New York Times,* "A Flower Market."

13. See Marx, *Machine in the Garden*; Slotkin, *Fatal Environment*. Ecocritics have increasingly rejected the notion that organic life is holistic, defined against the fracturing influence of technology and other human incursions. See, for example, Phillips, *Truth of Ecology*. New material studies go further in deconstructing the notion that the environment is an integral entity subject to human actions but ontologically distinct.

14. Marvel, "Mower against Gardens."

15. Rogers, *Writing the Garden.*

16. In Poe's story, the experience of Arnheim's fantastical domain is entirely navigated by a force beyond the visitor's control. Sitting in a boat, the viewer is pulled along a set course by an invisible current until a final orgiastic vista "bursts upon the view" with a "gush of entrancing melody." See Poe, "Domain of Arnheim," 869. Poe's story heightens attention to the common practice of creating "articulated landscapes." As Lynne Feeley describes, "An articulated landscape is one designed to move people through it in a particular order to produce particular views," and as she points out, such landscapes were especially common on plantations. See Feeley, "Swamps, Squash, Slavery."

17. Hawthorne, *American Notebooks*, 331. Subsequent references to this text appear parenthetically and refer to this edition.

18. Fuller, *Summer on the Lakes*, 93.

19. Stowe, "House and Home Papers," 576.

20. Philo Florist, "On the Cultivation of Flowers," 125.

21. As Erica Hannickel notes, the "morality of the scientific management of nature was hotly contested" (*Empire of Vines*, 146).

22. Hawthorne, *Selected Tales and Sketches*, 388. Subsequent references to this text appear parenthetically.

23. Brickhouse, "Hawthorne in the Americas."

24. The role of women gardeners has largely been placed within a women's history narrative that centers on domestic gardening activities. See Page and Smith, *Women, Literature, and the Domesticated Landscape*; Shteir, *Cultivating Women, Cultivating Science*; George, *Botany, Sexuality, and Women's Writing*. Male horticultural pursuits have by and large been placed in a separate historiography, despite the fact that women practitioners were often in communication with male institutional scientists (Baym, *American Women of Letters*; Gianquitto, "*Good Observers of Nature*").

25. Page and Smith, *Women, Literature, and the Domesticated Landscape*, 17.

26. Brickhouse, "Hawthorne in the Americas," 231.
27. In this vein the garden figures as "a horticultural experiment supervised by a strange aristocrat, a nation obsessed with the pedigree of its population, an operating theatre in which a woman dies, and a woman's uterus" (76). See Medoro, "'Looking into the Inmost Nature.'"
28. See chapter 4 on Stowe. See also Rosenthal, "Floral Counterdiscourse."
29. Lynch, "'Young Ladies Are Delicate Plants,'" 692, emphasis original.
30. Manning, *History of the Massachusetts Horticultural Society*, 266.
31. Hertz letter.
32. Unnamed newspaper, "Palm House," accessed through Kewensia, with curator's note that article likely came from a Glasgow publication.
33. Hawthorne, *Selected Tales and Sketches*, 391, 392. Subsequent references to this text appear parenthetically. As Amy M. King points out, the Victorian courtship plot is intimately tied to the scientific classification of flowers. See King, *Bloom*.
34. On sympathy, see Hendler, *Public Sentiments*, and Yao, *Disaffected*.
35. Hendler, *Public Sentiments*, 5.
36. Sophia Bamert uses Stacy Alaimo's concept of transcorporeality to describe how the text "offers a means of perceiving the material, embodied effects of racial formation." See Bamert, "Miasmas in Eden."
37. Korobkin, "Scarlet Letter of the Law," 193. As Korobkin argues, "Because community stability depends on each member's self-restraint, Hester's conformity to behavioral expectations helps hold the community together even when her thoughts may be at their bitterest. If protecting society from disruptive assaults is a paramount goal, then such a system helps preserve order, while leaving each individual the sanctum of his or her own mind. Whether that very limited freedom is enough is of course quite a different question" (201).
38. Sacvan Bercovitch describes the novel as "a story of socialization in which the point of socialization is not to conform, but to consent. Anyone can submit; the socialized believe. It is not enough to have the letter imposed; you have to do it yourself, and that involves the total self—past, present, and future; private and public; thought and passion." See *Office of the Scarlet Letter*, xiii. Along similar lines, Lauren Berlant describes Hawthorne's construction of a national symbolic that shows how "in America the possibility of a sutured national-popular consciousness has been a central distinguishing mark of state mythic utopianism. America has from the beginning appropriated the aura of the neutral territory, the world beyond political dissensus, for its own political legitimation." See Berlant, *Anatomy of National Fantasy*, 32.
39. Hawthorne, *Scarlet Letter*, 14. Subsequent references to this text appear parenthetically and refer to this edition.
40. Mann, *Lectures on Education*, 80.
41. *Youth's Companion*, "Variety: Poisonous Plant," emphasis original.
42. Brodhead, *Culture of Letters*.

43. Continuing the analogy, the author writes, "Agriculture makes the one productive, education the other. Brought under cultivation, the *soil* brings forth wheat and corn and good grass, while the weeds and briars and poisonous plants are all rooted out: so *mind* brought under cultivation, brings forth skill, and learning, and sound knowledge, and good principles: while ignorance and prejudice and bad passions, and evil habits, which are the weeds and briars and poisonous plants of the mind, are rooted out and destroyed." See Mc Vickar, "On Education," 87, emphasis original.

44. *Ladies' Companion*, "Picciola," 147.

45. *New-Yorker*, "Picciola," 67.

46. *Godey's Lady's Book*, "Picciola," 57.

47. "Flowers," June 15, 1867, reprinted from the *Christian Freeman*.

48. Tom tells Jim to "call it Pitchiola [*sic*]—that's its right name when it's in prison. And you want to water it with your tears."

49. Clifford Carruthers wagers it near impossible Hawthorne didn't read the novel, although he reads the rosebush as a symbol of exacting control. I want to suggest the opposite. See Carruthers, "'Povera Picciola' and *The Scarlet Letter*."

50. Bercovitch, *Office of the Scarlet Letter*, 18.

51. See chapter 4 for a fuller description of Topsy, Stowe, and horticulture.

52. Blackford, *Mockingbird Passing*, 119.

53. Hawthorne, *American Notebooks*, 176.

54. Weeds, as Michael Pollan notes, are "plants particularly well-adapted to man-made places. They don't grow in forests or prairies—in 'the wild' [but . . .] thrive in gardens, meadows, lawns, vacant lots, railroad sidings. . . . They grow where we live, in other words, and hardly anywhere else." See Pollan, "Weeds Are Us."

55. Laura Doyle, in registering the novel's process of Anglo-American colonization, focuses on Hester as the one who "re-nativizes" the land. See Doyle, *Freedom's Empire*, 328.

56. Scholars of nineteenth-century literary nationalism have characterized the "geographic fluidity" that is a hallmark of transnational and trans-American nationalist formulations. See Levine, *Dislocating Race and Nation*, 5; Doyle, *Freedom's Empire*; and Goudie, *Creole America*.

57. Edelman, *No Future*, 1–2.

58. Edelman, *No Future*, 3.

59. Rifkin, *Settler Common Sense*, 41.

60. Holly Jackson points out that "the best-selling racial scientific writings of the 1840s and 1850s returned to the anti-aristocratic rhetoric of inheritance reform to defend American slavery on the grounds that it reflected a natural inequality between the races, in opposition to the 'artificial' class distinctions under systems of hereditary aristocracy." See Jackson, "Transformation of American Property," 280.

61. Ritvo, "At the Edge of the Garden," 373.

62. *Horticulturalist and Journal of Rural Art and Rural Taste*, "Horticultural Festival at Faneuil Hall, Boston," 3, 5, 225.

63. Hawthorne, *Our Old Home*, 421. Subsequent references to this text appear parenthetically.

64. Pauly, *Fruits and Plains*, 7.

65. Pauly, *Fruits and Plains*, 53.

66. Hawthorne, *House of the Seven Gables*, 231, 87. Subsequent references to this text appear parenthetically.

67. *Genesee Farmer and Gardener's Journal*, "Action of Poison on Vegetable Structure," 4, 15.

68. Others argued that different plants drew different nutrients from the soil, thereby preventing rapid depletion of certain nutrients, and still others posited that "some other root crops . . . possess the property of adding to, rather than taking from, the quantity of vegetable matter in the soil" (170). *American Agriculturalist*, "Philosophy of the Rotation of Crops."

69. Pacheco, "'Vanished Scenes,'" 190.

70. In a different context, Whitman later meditates on similar issues in "This Compost." Thoreau likewise addresses soil history by asking rhetorically what we "have not written on the face of the earth already, clearing, and burning, and scratching, and harrowing, and ploughing, and subsoiling, in and in, and out and out, and over and over, again and again, erasing what they had already written for want of parchment." Thoreau, *Week on the Concord and Merrimack Rivers*, 9.

71. In explaining his decision to call the work a romance rather than a novel, Hawthorne states in the preface, "When romances do really teach anything, or produce any effective operation, it is usually through a far more subtile [*sic*] process than the ostensible one" (2).

72. Kaplan, *Anarchy of Empire*, 4.

73. Coale, "Romance of Mesmerism." For an extensive account of Hawthorne and mesmerism, see Ogden, *Credulity*.

74. Wood, "Burdock and the Violet."

75. *Putnam's Monthly Magazine*, "Chat about Plants," 433.

76. Consequent to these divisions, Timothy Mitchell notes, are several patterns in Western critical discourse: first, the interpretation of discrete events in relation to a universal that, as a category, is "founded within and expressed by the particular history of the West" (*Rule of Experts*, 29); and second, an understanding that humans are the only historical actors. Bruno Latour theorizes the power of "the moderns" in relation to a separation of, on the one hand, nature and culture and, on the other, human and nonhuman. "The critical power of the moderns lies in this double language," he writes. "They can mobilize Nature at the heart of social relationships, even as they leave Nature infinitely remote from human beings; they are free to make and unmake their society, even as they render its laws ineluctable, necessary, and absolute." See Latour, *We Have Never Been Modern*, 37. Leaving aside Latour's deliberately vague definition of "the moderns," his critique provides a helpful way of theorizing the stakes of mobilizing power along the interplay of nature, science, and society as discrete yet constantly hybridizing fields.

3. BOTANICAL AGENCY

1. See Altschuler, *Medical Imagination*.
2. Recognizing this fact helps us see how Dickinson's scientific language and her sentimental depictions of plants are not two entirely separate knowledge systems, and helps bring studies of her sentimentalism and her empiricism closer together. For instance, it complicates the position that "science for Dickinson represented not simply a positive field of learning but a challenge to every kind of sentimental domestic piety." See Giles, "'Earth Reversed Her Hemispheres,'" 7. A number of scholars have pointed out the importance of sentimentality to Dickinson's writing. See Noble, *Masochistic Pleasures of Sentimental Literature*; Stein, *Shifting the Ground*.
3. See Walls, *Seeing New Worlds*.
4. For discussions of Dickinson in a global perspective, see Gerhardt, *Place for Humility*; Giles, "'Earth Reversed Her Hemispheres'"; Miller, *Reading in Time*.
5. Dickinson, *Poems of Emily Dickinson*. All subsequent Dickinson poems are from this edition.
6. Gerhardt, "'Often Seen—but Seldom Felt,'" 73.
7. Farr, *Gardens of Emily Dickinson*, 120.
8. B.K. Bliss and Sons, *B.K. Bliss & Sons' Illustrated Spring Catalogue*, 8.
9. *Barr and Sugden's Guide*.
10. Dearborn, "Mass Horticultural Society."
11. See Farr, *Gardens of Emily Dickinson*, 120, on botanical catalogues Dickinson likely read; Breck, *Flower-Garden*, 28.
12. Farr, *Gardens of Emily Dickinson*, 110.
13. Gerhardt, "'Often Seen—but Seldom Felt,'" 68.
14. Bewell, "Romanticism and Colonial Natural History," 15.
15. Farr, *Gardens of Emily Dickinson*, 133.
16. For an excellent discussion of aesthetics and global botany, see Kelley, *Clandestine Marriage*.
17. Miller, *Reading in Time*, 125.
18. In a nice bit of irony, Dickinson's friend Thomas Wentworth Higginson retrospectively cast her in precisely this mold. In an 1894 essay in *The Chap-book*, Higginson excoriates American poets for using ornithological or botanical material that is not native to American soil: "It cost half a century of struggle for Lowell to get the bobolink and the oriel established in literature; and Emerson the chickadee, and Whittier the veery," he writes, adding, "At a later period, Emily Dickinson added the blue jay." Higginson melodramatically urges that "the literature of a nation must still have its own flowers beneath its feet, and its own birds above its head; or it will perish." Nowhere is Higginson's desire for fixity more apparent than in his appeal to "the genuine concrete earth" of America. Higginson, "A Step Backward?," 330D.
19. For discussions of Dickinson and Darwin, see Chow, "'Because I See—New Englandly'"; Eberwein, "Outgrowing Genesis"; Guthrie, "Darwinian Dickinson"; Peel, *Emily Dickinson and the Hill of Science*.

20. Darwin, *On the Origin of Species*, 308.
21. Thoreau, "The Succession of Forest Trees" (1860), in *"Wild Apples" and Other Natural History Essays*, 93.
22. See Beer, *Darwin's Plots*.
23. Mitchell, *Rule of Experts*, 10.
24. Historical Note, "Finding Aid for Emily Dickinson Botanical Specimens, Undated."
25. Historical Note, "Finding Aid for Emily Dickinson Botanical Specimens, Undated."
26. For a discussion of flowers and gift-based circulation in Dickinson's letters to local friends, see Crumbley, "Dickinson's Correspondence and the Politics of Gift-Based Circulation," and Tingley, "'Blossom[s] of the Brain.'"
27. Endersby, *Imperial Nature*, 107. See also chap. 4 in Susan Scott Parrish's *American Curiosity*.
28. For a discussion of distant correspondence in the nineteenth century, see Dierks, *In My Power*, 110.
29. Elizabeth Maddock Dillon has argued that we need to pay more attention to "a public sphere that maps onto the geopolitics of religion" (196) in addition to the dominant organization around nation-states. See Dillon, "Religion and Geopolitics in the New World."
30. Letter to Dr. and Mrs. J.G. Holland (L315), in Dickinson, *Letters of Emily Dickinson*, 168. Subsequent references to Dickinson's letters appear parenthetically and are from this collected volume.
31. Kelley, *Clandestine Marriage*, 3.
32. Kelley, *Clandestine Marriage*, 7. Kelley suggests that in this sense botany "offered a conduit to some of romanticism's most persistent inquiries, beginning with the nature of nature and of life and including the debate about whether nature or spirit should dominate, the global market of commodity plants, the relation between scientific inquiry and aesthetic pleasure, and the epistemological value accorded to concepts and particulars" (7).
33. See McDowell, *Emily Dickinson's Gardens*, 25 and Leslie Morris's introduction to the facsimile edition of the Herbarium.
34. Erasmus Darwin's 1789 botanical poem "The Loves of the Plants" helped popularize this association. For good discussions of the relationship between taxonomic systems and social values, see Browne, "Botany for Gentlemen" and Ritvo, *Platypus and the Mermaid*.
35. See Guthrie, "Darwinian Dickinson."
36. For an excellent discussion of Dickinson and the language of flowers as a form of private emotional symbolism, see chap. 5 in Petrino, *Emily Dickinson and Her Contemporaries*.
37. Linnaeus, *Philosphia Botanica*, 9; Bartram, *Travels*, liv.
38. Bartram, *Travels*, liv.
39. Gaudio, "Elements of Botanical Art," 437.

40. Darwin and Darwin, *Power of Movement in Plants*, 573.
41. De Candolle, *Elements of the Philosophy of Plants*, 238.
42. *Literary World*, "How Plants Behave," 114.
43. E.H.C., "Gleanings," 148. This article was excerpted in an 1870 *Ohio Farmer* piece called "Sensibility of Nature," which examines nature that "could not talk words, but . . . could talk things" and proposes a study of "the wits . . . that a rosebush has" (551).
44. *All the Year Round*, "Have Plants Intelligence?," 3.
45. *Youth's Companion*, "Sleep of Plants," 108.
46. *Appleton's Journal of Literature, Science and Art*, "Soul of Plants," 397.
47. Allewaert, *Ariel's Ecology*, 55.
48. Walls, *Seeing New Worlds*, 56.
49. Thoreau, *Wild Fruits*, 242. For more on Thoreau's understanding of life, see Arsić, *Bird Relics*.
50. *Eclectic Magazine of Foreign Literature*, "Can We Separate Animals from Plants?," 608.
51. *Eclectic Magazine of Foreign Literature*, "Can We Separate Animals from Plants?," 614.
52. See Ritvo, *Platypus and the Mermaid*.
53. Phelps, *Familiar Lectures on Botany* (1854), 241.
54. Hitchcock, "Edward Hitchcock Classroom Lecture Notes, 'Botany.'"
55. Saintine, *Picciola*, 55, 51.
56. Saintine, *Picciola*, 72.
57. Farr, *Gardens of Emily Dickinson*, 186.
58. For a discussion of this poem in the context of the Civil War, see Marrs, *Nineteenth-Century American Literature and the Long Civil War*, 126. Cristanne Miller also speculates that Dickinson references the war in this poem. See Miller, *Reading in Time*, 160. Eliza Richards has more generally explored the ways in which Dickinson grappled with Civil War news and remote suffering and "the difference between the unknowable experience of trauma and the vicarious imaginings of that experience inspired by reading about it" (165). See Richards, "'How News Must Feel When Traveling.'" See also Favret, *War at a Distance*.
59. "'Nature' Is What We See" (F721 B) similarly illustrates this point. The speaker tests three theses on nature that hinge on human perception: seeing, hearing and knowing. In the first two cases, the speaker begins with a diverse list of what nature "is" before offering a more totalizing claim: for instance, in the first stanza, "'Nature' is what We see— / The Hill—the Afternoon— / Squirrel—Eclipse—the Bumble bee— / Nay—Nature is Heaven—." This list points out the discrete entities that can stand metonymically for Nature.
60. See Berlant, *Female Complaint*.
61. See Dillon, "Sentimental Aesthetics"; Howard, "What Is Sentimentality?"
62. Dillon, "Sentimental Aesthetics," 498.
63. See Latour, *We Have Never Been Modern*.

64. Howell, "In the Realm of Sensibility," 408.
65. Dillon describes how the history of aesthetics "developed in response to the revolutions of the eighteenth century that ushered in liberal political regimes and societies oriented around (newly) autonomous, self-governing citizen-subjects" ("Sentimental Aesthetics," 497).
66. Folsom and Price, "Dickinson, Slavery, and the San Domingo Movement."
67. Eduardo Kohn's *How Forests Think* and Anna Tsing's *Mushroom at the End of the World* are prominent exceptions in their focus on forest ecologies and mushrooms. See Kohn, *How Forests Think* and Tsing, *Mushroom at the End of the World*. For more multispecies ethnographies, see Kirksey, *Emergent Ecologies*; Kirksey, Schuetze, and Helmreich, *Multispecies Salon*; Scaramelli, *How to Make a Wetland*. In literary studies, see also recent work by Stacy Alaimo, Dana Luciano, and Geoffrey Sanborne. Natania Meeker and Antónia Szabari's *Radical Botany* shows how integral plants have been to imagining new worlds. Plant studies is gaining critical traction, as evidenced by the work of figures like Robin Kimmerer, Michael Marder, and Catriona Sandilands and recent edited collections such as Gagliano, Ryan, and Vieira, *Language of Plants*, and Laist, *Plants and Literature*.

4. BOTANICAL ABOLITIONISM

1. Stowe, "Meditations from Our Garden Seat," 349.
2. Frederick Douglass, reprinted "Meditations from Our Garden Seat" in *Frederick Douglass' Paper* later the same month.
3. Martha Schoolman convincingly reads Stowe's depiction of the maroon community as pointing to a "counter tradition of revolt and resistance that she sees as having the potential to contest the practice of conservative compromise that had come to characterize deliberative democracy in the United States" (*Abolitionist Geographies*, 169). See also Duquette, "Republican Mammy?"
4. On this, see Bolker, "Stowe's Birds."
5. Stowe, "Harriet Beecher Stowe as a Mother," 223.
6. Hedrick, *Harriet Beecher Stowe*, 126, 154.
7. *Family Magazine*, "Botany," 52.
8. For a further discussion of the history of antebellum amateur botany, see Keeney, *Botanizers*.
9. Baym, *American Women of Letters*, 5.
10. See Beecher, *Treatise on Domestic Economy*, 252.
11. Ward, *On the Growth of Plants*, 87, 89, emphasis original.
12. Stowe, *Sunny Memories of Foreign Lands*, 249, 247.
13. Stowe, *Oldtown Folks*, 103.
14. Phelps, *Familiar Lectures on Botany*, 12.
15. This mode of thinking had strong moral undertones; as Amanda Anderson, George Levine, and Colin Jager have argued, scientific epistemology was firmly linked to ethical positioning as it worked in tandem with religion. See Anderson, *Powers of Distance*; Jager, *Book of God*; Levine, *Dying to Know*, 2.

16. Johnson, "Rasselas: A Tale" (1759), 222.
17. Walls, *Seeing New Worlds*, 78.
18. For a discussion of the relationship between floral tropes and domestic ideology, see Tina Gianquitto, *"Good Observers of Nature"*; Seaton, *Language of Flowers*.
19. Schuller, *Biopolitics of Feeling*, 2.
20. Allewaert, *Ariel's Ecology*, 18.
21. Bolker, "Stowe's Birds," 239.
22. Fuller, "Autobiographical Romance," 32. Steele dates the writing of "Autobiographical Romance" to 1840.
23. Phelps, *Story of Avis*, 90, 91, 209.
24. Recent work in literary history and the history of science has done much to debunk the still common correlation of nineteenth-century science with unemotional objectivity. To the contrary, as these critics show, emotion and imagination were central to scientific practice. Jessica Riskin identifies the way in which eighteenth-century empiricism, long associated with detached observation and the strict disavowal of emotion, relied upon the sentimental processing of empirical sensations. See Riskin, *Science in the Age of Sensibility*. Sari Altschuler points out that as medical practitioners in early America slowly and unevenly pivoted from rationalism toward empiricism, they experimented with literary forms and imaginative methods to grapple with medical problems. See Altschuler, *Medical Imagination*. In a related vein, Amanda Jo Goldstein considers the Romantic poetry that stood "as a privileged technique of empirical inquiry: a knowledgeable practice whose *figurative* work brought it closer to, not farther from, the physical nature of things" (7). See Goldstein, *Sweet Science*. Emily Ogden and Jonathan Tresch show how enchantment remained alive in the scientific and technological developments of the early nineteenth century, marshaled to secular and administrative ends. See Ogden, *Credulity*, and Tresch, *Romantic Machine*.
25. See Brown, *Domestic Individualism*, 8; Merish, *Sentimental Materialism*.
26. Page and Smith, *Women, Literature, and the Domesticated Landscape*, 8.
27. Stowe, "Harriet Beecher Stowe to Frederick Law Olmsted."
28. Stowe, *Uncle Tom's Cabin*, 276.
29. For an expanded discussion of the role of horticulture to Revolutionary era politicians, see Wulf, *Founding Gardeners*.
30. See Allewaert, "Swamp Sublime," 341.
31. See *North American Review*, "Botany of the United States," 134.
32. John Conron describes how in the American picturesque, "Wildness . . . is defined by the character of its resistance to an inherent geometry." See Conron, *American Picturesque*, 34.
33. Brontë, *Jane Eyre*, 373.
34. As Maria Karafilis notes, the swamp is a kind of heterotopia where the "logic of dividing, parceling, and possessing . . . is not applicable." See "Spaces of Democracy," 29.
35. Aptheker, "Marooners within the Present Limits," 152.

36. Ball, *Slavery in the United States*, 166.
37. *Waddill v. Martin*, 38 NC 562 (1845).
38. *State v. Mann*, 13 NC 263 (1829).
39. Helper, *Impending Crisis of the South*, 133.
40. Stowe was moved by Olmsted's assessments of southern culture, writing to him, "I am charmed with your book—It is extremely graphic and readable and exceedingly calculated to do good—I hope it will circulate and be extensively read at the South—I think I never saw a work on the subject calculated to do more good with less friction." Letter to Olmsted, June 1856. Olmsted attributes "exhausted fields" in Virginia to a "laisser faire [*sic*] principle [that] reigned in their politics," allowing for soil depletion through the overplanting of tobacco. See Olmsted, *Journey in the Seaboard Slave States*, 241. Olmsted similarly writes earlier that "the climate [in the South] would permit the culture of many valuable plants which cannot exist at New-York, and the indigenous productions of the soil are much more varied. But little advantage has been taken of this opportunity; the productions of agriculture being less-varied in general than the free states. The difficulty of introducing anything new into the routine of labor where slaves are employed, is the first reason for this." See Olmsted, "Letters on the Productions, Industry and Resources of the Southern States," 128.
41. Kenrick, *Farmer's Register*, 235. The article was first published in the *Farmer's Cabinet* on April 30, 1839, then reprinted in the *Liberator* in July of the same year, and again in the *New England Farmer, and Horticultural Register* two weeks after that.
42. For more on botanical taxonomy and early Americans' vocabulary for human sexuality, especially as tied to race, see LaFleur, *Natural History of Sexuality*.
43. Bennett's *Being Property Once Myself* engages how Black authors in the United States have fought this correlation and used animal figures to theorize Black sociality.
44. McCune Smith, "Introduction."
45. Kingbury, *Hybrid*, 93.
46. Lindley, "Remarks on Hybridising Plants," 117.
47. Lindley, "Remarks on Hybridising Plants," 117.
48. Jackson-Houlston, "'Queen Lilies'?," 85.
49. Lemire, "*Miscegenation*," 39.
50. *Eclectic Magazine of Foreign Literature*, "Pleasure of Botany and Gardening," 117.
51. Otter, "Stowe and Race," 31.
52. As one dissident defines it, "the proneness to consider successive phenomena as connected with each other by the relations of cause and effect—when they have been entirely distinct—and their association altogether incidental." W.D., "Transmutation of Plants," 257.
53. W.D., "Transmutation of Plants," 257, emphasis original.
54. Finseth, *Shades of Green*, 8.
55. Pauly, *Fruits and Plains*, 32, 69.

56. The writer describes the *Victoria Regia* as "an aquatic plant, which demands the atmospheric temperature of the equator, and at least twenty or thirty feet of space to extend its leaves, which requires to be grown in a pond of water, kept to the temperature of 85 degrees Fahrenheit, and still more, to have this water gently agitated, to imitate the movement of a stream." See *Horticulturalist and Journal of Rural Art and Rural Taste*, "New Water Lily—Victoria Regia," 275.

57. Amy Kaplan has described how "the home is a mobile space that became in one way encompassed and in another expelled the foreign within" (Kaplan, *Anarchy of Empire*, 19).

58. Thoreau, "The Succession of Forest Trees" (1860), in *"Wild Apples" and Other Natural History Essays*, 94.

5. BOTANICAL SOCIETIES

1. Douglass, "Pumpkins." It is worth noting here that Douglass is likely referring to pumpkins raised by his wife, Anna Murray Douglass, who was known to be an exceptional gardener. To make the larger point about racialized labor, Douglass's use of the collective "our" here conceals the gendered dynamic of household labor. Douglass's travel schedule during this period likely precluded the kind of careful attention to gardening he developed later in life. See McFeely, *Frederick Douglass*, 82, 154.

2. Carlyle, "Occasional Discourse on the Negro Question."

3. For more on the way whites used nature as the basis for claims of racial difference, see Finseth, *Shades of Green*; Outka, *Race and Nature*.

4. Scholars have analyzed Booker T. Washington and W. E. B. DuBois's divergent perspectives on cultivation's liberatory potential. See Posmentier's excellent work in *Cultivation and Catastrophe*. John Claborn also notes Zora Neale Hurston's vision of a Black-owned agricultural cooperative. See Claborn, *Civil Rights and the Environment in African-American Literature*, 115.

5. To quote Smith, "We usually think of 'environmental thought' as a set of ideas and arguments aimed at preserving the wilderness and maintaining a viable ecosystem, and defined by or tending toward ecocentric values. Under that definition there is little in the black political tradition that qualifies." See Smith, *African American Environmental Thought*, 3. Ramachandra Guha and Joan Martinez-Alier have made a similar argument about the exclusions of "full-bellied" environmentalism. See Guha and Martinez-Alier, *Varieties of Environmentalism*. The canon of recognized environmental thought is widening, as Smith and Guha's critical work attests, along with recent work by Karen Kilcup, Sonya Posmentier, Rob Nixon, and others.

6. On this pastoralism, conservation, and white supremacy, see also Alexandre, *Properties of Violence*; Wagner-McCoy, "Virgilian Chesnutt"; and Ray, "Risking Bodies in the Wild."

7. Holbrook, "Democracy of Science." As Emily Pawley points out, "During the mid-1840s a coalition of chemists, agricultural journalists, and improving agricultur-

alists strongly promoted quantitative chemical analysis as a way of identifying monetary value in the material transactions of the farm." See Pawley, "Accounting with the Fields," 463.

8. Posmentier, *Cultivation and Catastrophe*, 29.
9. Sutter, "World with Us." Donna Haraway and others have used the term "plantationocene" as shorthand to address the effects of plantation agriculture systems on the environment and disenfranchised labor. See Haraway et al., "Anthropologists Are Talking about the Anthropocene."
10. As we saw in chapter 4, Stowe was drawn to the power of this critique. For a discussion of soil exhaustion as the basis for antislavery arguments, see also Cristin Ellis, "Amoral Abolitionism."
11. James, "'Buried in Guano.'"
12. Historians of slavery, agriculture, and capitalism have all traced the ways in which planters cohered around scientific management as a tool of redress for exhaustion. See Pawley, "Accounting with the Fields"; Olmstead and Rhode, *Creating Abundance*. The belief that technological innovation could mitigate destructive practices shares parallels with some veins of techno-optimism in the climate change debate today.
13. I take this point from Benjamin Cohen, who has argued that the Agricultural Society of Virginia was "not simply groups of men who wanted better farming for farming's sake; they were powerful political entities that sought to control the sanctity and the future of the Southern plantation system and, as such, were overtly political" (143). See Cohen, *Notes from the Ground*.
14. Rosenthal, "Slavery's Scientific Management."
15. Ellis, "Amoral Abolitionism," 287.
16. Douglass, *Narrative of the Life of Frederick Douglass*, 29. Subsequent references to this text appear parenthetically and refer to this edition.
17. Critics have attentively illustrated the power of signification in this scene. Edward Dupuy reads Lloyd's garden as embodying a "myth of paradise" for Colonel Lloyd, his "claim to Eden . . . signif[ying] his own riches and the riches of the land—his regeneration" (28). See Dupuy, "Linguistic Mastery and the Garden of the Chattel." Michael Bennett reads the biblical typology of both the garden and the tar. See Bennett, "Anti-pastoralism, Frederick Douglass, and the Nature of Slavery," 200.
18. Drayton, *Nature's Government*, 45.
19. Ian Finseth calls this moment "antigeorgic" since Douglass shows how "the plantation is not self-sufficient but parasitic" (*Shades of Green*, 280).
20. Douglass, *My Bondage and My Freedom*.
21. As Myra B. Young Armstead has noted, the labor histories of enslaved and paid professional gardeners are often occluded in the historiography of horticulture and scientific agriculture, diminishing our knowledge of those who carry out much of the material work. Armstead also describes the coveted status of Scottish gardeners during the antebellum period. See Armstead, *Freedom's Gardener*.
22. See Pruitt, "Reordering the Landscape."

23. Douglass, *My Bondage*, 82.
24. This contrasts sharply with the ways that Douglass boasts of his grandmother's farming expertise, which was known throughout the region. As Ellis points out, Douglass depicts the way his grandmother leverages her agricultural expertise as that of a "canny capitalist" (Ellis, "Amoral Abolitionism," 286). But her financial profit is also the result of her "high reputation" that comes from distributing her expertise to neighbors within the community.
25. Ellis, "Amoral Abolitionism," 277. Ellis also powerfully reminds us of the importance of environmental crisis "in shaping antislavery politics in America."
26. Carr, "Southern Planter."
27. South Side, "For the Southern Planter," 204.
28. Youmans, "Chemistry Applied to the Mechanic and Farmer," 206.
29. Douglass, "Lecture on Our Nation's Capital."
30. For an in-depth discussion of this alimentary and economic system in Georgia, see Stewart, *"What Nature Suffers to Groe."* As Stewart, Judith Carney, and Sharla Fett have illustrated, there is little extensive knowledge about these patches, and gaps in the archives remain vast. See Carney and Rosomoff, *In the Shadow of Slavery*; Fett, *Working Cures.*
31. Quarles, *Black Abolitionists*, 65.
32. Northup, *Twelve Years a Slave*, 97. Subsequent references to this text appear parenthetically and refer to this edition.
33. Jacobs, *Incidents in the Life of a Slave Girl*, 47, 48.
34. Jacobs, *Incidents in the Life of a Slave Girl*, 56, 57.
35. Schoolman, "Violent Places."
36. Charles Baraw, 454. Baraw borrows the term "fugitive tourism" from John Ernest's *Liberation Historiographies.*
37. Schoolman, "Violent Places," 7.
38. Barry, *Fruit Garden*, iii.
39. Ron, "Summoning the State," 349.
40. Rusert, *Fugitive Science*, 7.
41. Tobin, *Colonizing Nature*, 174.
42. For more on the circulation and reception history of the *Christian Recorder*, see Eric Gardner, *Black Print Unbound.*
43. Thomson, "Speech of George Thompson," 3. Subsequent references to this text appear parenthetically.
44. *Punch*, "Black Statue to Thomas Carlyle."
45. After Benedict Anderson's work on imagined communities, a number of scholars have expanded our understanding of print culture. See Brodhead, *Culture of Letters*; Loughran, *Republic in Print*; Ernest, *Nation within a Nation*; Fagan, *Black Newspaper and the Chosen Nation*; Spires, *Practice of Citizenship.*
46. Gardner, *Black Print Unbound*, 14.
47. As Daegan Miller notes, references to these lands appear no fewer than eighteen times in *The North Star*. See *This Radical Land.*

48. Miller, *This Radical Land*, 62.
49. Miller describes the "competing, aristocratic, largely Southern agrarianism" that espoused "leisure—the leisure to think and write and cultivate the finer aspects of Western culture—not labor" as the basis for America's successful democratic experiment (*This Radical Land*, 58).
50. Miller, *This Radical Land*, 60.
51. F.D., "Smith Lands."
52. Greeley, "Farm Life."
53. *Frederick Douglass' Paper*, "Make Your Sons Mechanics and Farmers."
54. Douglass, "Letter to Harriet Beecher Stowe," 216.
55. Douglass, *Selected Speeches*, 216. While the project failed to get off the ground, Stowe and Douglass shared a belief in "an honest earth" that might offer a return for an honest labor system. Stowe's "Meditations from Our Garden Seat," discussed in chapter 4, appeared in *Frederick Douglass' Paper* on August 31, 1855. In that essay, Stowe links plant cultivation and personal cultivation.
56. *Frederick Douglass' Paper*, "The Word 'White.'"
57. Woodson, "Gardening."
58. A.G.B., "For Frederick Douglass' Paper."
59. *Provincial Freeman*, "Baby Show in Ohio."
60. *Albany Cultivator for April*, "Farmer's Garden."
61. The author "W.B." was likely the horticulturalist William Bacon, who became the editor at *Berkshire Farmer*. The editor at the *New England Farmer* surmised this in the August 24, 1842, issue of the periodical (62).
62. *Frederick Douglass' Paper*, "Colored People's Agricultural Fair at Geneva."
63. *Frederick Douglass' Paper*, "He Who Is a Good Theoretical Agriculturalist. . . ."
64. *Albany Cultivator*, "Jardin des Plantes."
65. Douglass, "Address Delivered." Subsequent references to this text appear parenthetically.
66. Eric Foner describes how despite the ways whites blocked their efforts toward landownership, newly emancipated Black people "establish[ed] as much independence as possible in their working lives, consolidate[d] their families and communities, and stake[d] a claim to equal citizenship" (Foner, *Reconstruction*, preface).
67. LeMenager, *Living Oil*.
68. James, "'Buried in Guano,'" 118.
69. Arthur Riss has argued that Douglass's "initiation into slavery [is] inseparable from his initiation into sentimentality. . . . Slavery violates the principles of sympathy and love; it separates families; disrupts the home, and victimizes children. Indeed, it is precisely such affective language that creates the sympathetic bond between Douglass and his readers central to sentimentality." See Riss, "Sentimental Douglass," 104. See also Foreman, "Sentimental Abolition in Douglass' Decade."
70. McClarin, "Cedar Hill."

CONCLUSION

1. Schiebinger, *Plants and Empire*, 5.
2. To take one example, the U.S. maple syrup industry is moving north into Canada as erratic spring conditions in New England limit the run of sap.
3. Tribal rights, including off-reservation rights, are likewise affected as species migrate and habitats change. This is one of many ways in which climate change amplifies the legacy of colonial violence for indigenous communities. In Whyte's words, "climate injustice is part of a cyclical history situated within the larger struggle of anthropogenic environmental change catalyzed by colonialism, industrialism and capitalism." See Whyte, "Is It Colonial Déjà Vu?" Rosalyn LaPier has also described how conservation projects in the United States, such as national parks, have restricted Native rights to gather culturally significant plants. Listen to her interview about fortress conservation on the NPR podcast *Inside/Out*.
4. Kimmerer, *Braiding Sweetgrass*, 57–58.
5. Pollan, "Intelligent Plant."
6. Pollan, "Intelligent Plant."
7. In a memorable scene in the 1999 romantic comedy *Notting Hill* this idea is played for laughs by taking it to the extreme. At a dinner party among friends, the protagonist Will is set up on a blind date with a woman named Keziah, who turns down a cooked dish because she is a "fruitarian." When pressed for more details, she responds, "We believe that fruits and vegetables have feeling so we think cooking is cruel. We only eat things that have actually fallen off a tree or bush—that are, in fact, dead already." The rest of the dialogue unfolds as follows: William: "Right. Right. Interesting stuff. So, these carrots . . ." Keziah: "Have been murdered, yes." William: "Murdered? Poor old carrots. . . . That's beastly!" Clearly plant sentience is not a serious proposition in contemporary popular culture.
8. Myers, "Conversations on Plant Sensing."
9. Jahren, *Lab Girl*, 279.
10. Wohlleben, *Hidden Life of Trees*, 16.
11. Wohlleben, *Hidden Life of Trees*, xiv.
12. Her own recent memoir likewise stresses these connections. See Simard, *Finding the Mother Tree*.
13. Jabr, "Social Life of Forests."
14. Powers, *Overstory*, 4, emphasis original.
15. Hamner, "Here's to Unsuicide."
16. Indeed, species-level thinking has been aligned with the universalism of Western colonization, as postcolonial theorists have illustrated, and there is a robust literature critiquing the Anthropocene for this reason. As Elizabeth DeLoughrey and Kyle Whyte both note, the concept of the Anthropocene often elevates the idea that we are facing an unprecedented rupture with the past rather than a disaster that has been continuously made and remade since the era of colonialism. Such a species-level criticism also evenly distributes responsibility for disasters that have

been driven by some. DeLoughrey refers to much Anthropocene talk as a "globalization discourse that misses the globe" (*Allegories of the Anthropocene*, 2).

17. Margulis first published on this idea in 1967. Sagan, "On the Origin of Mitosing
Cells."

18. Jabr, "Social Life of Forests"; Favini, "What if Competition Isn't as 'Natural' as We
Think?"

19. Efforts are ongoing around the globe to extend rights to nature. In 1972 law
professor Christopher Stone wrote a slim volume called *Should Trees Have Standing?* in response to a proposed Disney development in the Sierra Nevada, which
was challenged in court by the Sierra Club. Stone argued that trees should be
granted legal standing and that the environment should be brought into the legal
system as a "holder of legal rights." Why is it, Stone asks, that corporate bodies,
nation-states, trusts, and other nonhuman entities have long held statutory and
constitutional rights, whereas rights for environmental entities have remained
inconceivable? Stone writes that "throughout legal history, each successive extension of rights to some new entity has been, theretofore, a bit unthinkable. We are
inclined to suppose the rightlessness of rightless 'things' to be a decree of Nature,
not a legal convention acting in support of some status quo." See Stone, *Should
Trees Have Standing?*, 6. Stone traces the direct connection between imagining
nature as property and the replication of a status quo that fails to protect nonhuman species from our actions. While the Supreme Court rejected the lawsuit,
Justice William Douglas issued a dissent arguing that environmental "objects"
should be allowed to sue for preservation. More recent cases have gained some
traction. In 2006 community members in Tamaqua, Pennsylvania, successfully
passed an ordinance that bestowed legal rights to nature. Since then a number of
legal challenges in the United States and abroad have helped lay the groundwork
for thinking about the natural world as an entity with legal rights and protections.
Ecuador made global news in 2008 when it nominally granted rights to Nature
in its constitution, although by most accounts rampant corruption has prevented
the realization of this constitutional promise. See Revkin, "Ecuador Constitution
Grants Rights to Nature." India, Colombia, Mexico, and Australia are a few of the
other countries at work on establishing similar protections.

20. Miller, "Reading Trees in Nature's Nation."

21. Hindi, "Fuck Your Lecture on Craft, My People Are Dying."

22. A number of scholars who study race have critiqued this form of green thinking.
For more on racialized ideas of "the human," see Weheliye, *Habeas Viscus*; Farooq,
Undisciplined; Wynter, "Unsettling the Coloniality of Being/Power/Truth/Free-
dom."

23. Kimmerer, *Braiding Sweetgrass*, 190.

24. Ruth Wilson Gilmore stresses the importance of understanding racial capitalism
as a relation and emphasizes the possibilities for meaningful change in cultivating place-based relations built on continually remade solidarities outside this
system. Gilmore advances the urgent need to change everything from the ground

up. Settler scholars, as Stephanie LeMenager characterizes most non-Indigenous folks working in the academy, have to be especially cognizant in the ways that they incorporate Indigenous teachings into their (our) own work so as to avoid appropriation. See LeMenager, "Love and Theft."

25. Susan Fraiman draws our attention to domesticity as the ordinary and the everyday. See *Extreme Domesticity*.

26. The existence of tropical houseplants in northern climates, as Beth Tobin has noted, is the result of eighteenth-century colonization (*Colonizing Nature*, xiii).

27. Plantkween Instagram post, November 6, 2020.

28. Johnson, "About."

29. The conservation movement that arose in the late nineteenth century pushed legal regulation of the landscape and consolidated federal power over lands deemed important for the public good. As environmental historian Karl Jacoby notes, "To create new laws also meant to create new crimes" (*Crimes Against Nature*, 2). From the passing of the Morrill Act and founding of the Department of Agriculture in 1862 to the rise of the national park system (Yosemite was protected in 1864, and Yellowstone became the first national park in 1872), the government increasingly regulated land use in ways that benefitted those in power. Plant introduction was systematized in 1897.

30. Anthropologist Charis Boke emphasizes how "Indigenous and local knowledges about plant medicines have historically been erased and re-transcribed as what is now called 'science' through colonization's global power dynamics." Boke, "Re-grounding Practice, Unsettling Knowledge."

31. Abena Dove Osseo-Asare powerfully describes the contests over power in the ways that contemporary pharmaceutical companies have sought new therapeutic plants in postcolonial Africa. See Osseo-Asare, *Bitter Roots*.

32. Purdy, *After Nature*, 7.

BIBLIOGRAPHY

Abumrad, Jad, and Robert Krulwich, hosts. "Smartyplants." *Radiolab*, February 13, 2018.

A.G.B. "For Frederick Douglass' Paper. From Our New Haven Correspondent." *Frederick Douglass' Paper*, October 19, 1855.

Ahsley, John M. "The Scientific History of a Plant." *Horticulturalist and Journal of Rural Art and Rural Taste*, June 1, 1851.

Alaimo, Stacy. *Bodily Natures: Science, Environment and the Material Self*. Bloomington: Indiana University Press, 2010.

Albany Cultivator. "The Jardin des Plantes." Reprinted in *Frederick Douglass' Paper*, February 19, 1852.

Albany Cultivator for April. "The Farmer's Garden." Reprinted in *Colored American*, July 28, 1838.

Alexandre, Sandy. *The Properties of Violence: Claims to Ownership in Representations of Lynching*. Jackson: University Press of Mississippi, 2012.

Allewaert, Monique. *Ariel's Ecology: Plantations, Personhood, and Colonialism in the American Tropics*. Minneapolis: University of Minnesota Press, 2013.

———. "Swamp Sublime: Ecologies of Resistance in the American Plantation Zone." *PMLA* 123 (2008): 340–357.

All the Year Round. "Have Plants Intelligence?" Reprinted in *The Independent*, June 30, 1870.

Altschuler, Sari. *The Medical Imagination: Literature and Health in the Early United States*. Philadelphia: University of Pennsylvania Press, 2018.

Ambrosoli, Mauro. "From Europe to the Americas: The Diffusion of Medicago sp., 16th–19th Centuries." *Histoire & Societes Rurales* 42, no. 2 (2014): 43–66.

American Agriculturalist. "Philosophy of the Rotation of Crops." June 1848.

———. "Visit to the Royal Gardens at Kew." Reprinted in *Friends' Review; a Religious, Literary, and Miscellaneous Journal* 13, no. 2 (April 21, 1860).

American Farmer, and Spirit of the Agricultural Journals of the Day. "On Guano as a Fertilizer." November 13, 1844.

Anderson, Amanda. *The Powers of Distance: Cosmopolitanism and the Cultivation of Detachment*. Princeton: Princeton University Press, 2001.

Anderson, Benedict. 1983. *Imagined Communities*. New York: Verso, 1991.

Appleton's Journal of Literature, Science and Art. "The Soul of Plants." 13 (June 26, 1869): 397.

Aptheker, Herbert. "Marooners within the Present Limits of the United States." In *Maroon Communities*, 3rd ed., edited by Richard Price. Baltimore: Johns Hopkins University Press, 1996.

Armbruster, Karla, and Kathleen Wallace, eds. *Beyond Nature Writing: Expanding the Boundaries of Ecocriticism*. Charlottesville: University of Virginia Press, 2001.

Armstead, Myra B. Young. *Freedom's Gardener: James F. Brown, Horticulture and the Hudson Valley in Antebellum America*. New York: New York University Press, 2012.

Armstrong, Nancy. *Desire and Domestic Fiction: A Political History of the Family Novel*. New York: Oxford University Press, 1987.

Arsić, Branch. *Bird Relics: Grief and Vitalism in Thoreau*. Cambridge, MA: Harvard University Press, 2016.

Ball, Charles. *Slavery in the United States: A Narrative of the Life and Adventures of Charles Ball, a Black Man*. New York: John S. Taylor, 1836.

Bamert, Sophia. "Miasmas in Eden: Atmosphere, History, and Narrative in *The House of the Seven Gables* and 'Rappaccini's Daughter.'" *Nathaniel Hawthorne Review* 43, no. 2 (Fall 2017): 1–18.

Barad, Karen. *Meeting the Universe Halfway: Quantum Physics and the Entanglement of Matter and Meaning*. Durham, NC: Duke University Press, 2007.

Baraw, Charles. "William Wells Brown, *Three Years in Europe*, and Fugitive Tourism." *African American Review* 44, no. 3 (Fall 2011): 453–470.

Barr and Sugden's Guide to the Flower Garden, &c. London, 1862. Royal Horticultural Society Library.

Barry, P. "A Visit to Kew, the English National Garden." *Horticulturalist and Journal of Rural Art and Rural Taste* 3, no. 10 (April 1849).

Barry, Patrick. *The Fruit Garden: A Treatise*. New York: Scribner, 1851.

Barton, Benjamin Smith. *Elements of Botany; or, Outlines of the Natural History of Vegetables*. London: J. Johnson, 1804.

Bartram, William. *The Travels of William Bartram*. Edited by Francis Harper. Athens: University of Georgia Press, 1998.

Batsaki, Yota, Sarah Burke Cahalan, and Anatole Tchikine, eds. *The Botany of Empire in the Long Eighteenth Century*. Washington, DC: Dumbarton Oaks, 2016.

Baucom, Ian. *History 4 Celsius: Search for a Method in the Age of the Anthropocene*. Durham, NC: Duke University Press, 2020.

Baym, Nina. *American Women of Letters and the Nineteenth-Century Sciences*. New Brunswick, NJ: Rutgers University Press, 2002.

Beam, Dorri. *Style, Gender, and Fantasy in Nineteenth-Century American Women's Writing*. New York: Cambridge University Press, 2010.

Beecher, Catharine. *A Treatise on Domestic Economy*. 1841. Boston: T. H. Webb, 1842.

Beecher, Catharine, and Harriet Beecher Stowe. *The American Woman's Home; or, Principles of Domestic Science*. New York: J.B. Ford, 1869.

Beer, Gillian. *Darwin's Plots: Evolutionary Narrative in Darwin, George Eliot and Nineteenth-Century Fiction*. Cambridge: Cambridge University Press, 1983.

Benfey, Christopher. *A Summer of Hummingbirds: Love, Art, and Scandal in the Intersecting Worlds of Emily Dickinson, Mark Twain, Harriet Beecher Stowe and Martin Johnson Heade.* New York: Penguin, 2008.

Bennett, Jane. *Vibrant Matter: A Political Ecology of Things.* Durham, NC: Duke University Press, 2010.

Bennett, Joshua. *Being Property Once Myself: Blackness and the End of Man.* Cambridge, MA: Harvard University Press, 2020.

Bennett, Michael. "Anti-pastoralism, Frederick Douglass, and the Nature of Slavery." In Armbruster and Wallace, *Beyond Nature Writing,* 195–210.

Bercovitch, Sacvan. *The Office of the Scarlet Letter.* Baltimore: Johns Hopkins University Press, 1991.

Berlant, Lauren. *The Anatomy of National Fantasy: Hawthorne, Utopia, and Everyday Life.* Chicago: University of Chicago Press, 1991.

———. *The Female Complaint: The Unfinished Business of Sentimentality in American Culture.* Durham, NC: Duke University Press, 2008.

Bewell, Alan. *Natures in Translation: Romanticism and Colonial Natural History.* Baltimore: Johns Hopkins University Press, 2017.

———. "Romanticism and Colonial Natural History." *Studies in Romanticism* 43, no. 1 (Spring 2004): 5–34.

B.K. Bliss and Sons. *B.K. Bliss & Sons' Illustrated Spring Catalogue and Amateur Guide to the Flower and Kitchen Garden.* New York, 1870.

Blackford, Holly. *Mockingbird Passing: Closeted Traditions and Sexual Curiosities in Harper Lee's Novel.* Knoxville: University of Tennessee Press, 2011.

Blackwell, Mark, ed. *The Secret Life of Things: Animals, Objects, and It-Narratives in Eighteenth-Century England.* Lewisburg, PA: Bucknell University Press, 2007.

Bleichmar, Daniela. *Visible Empire: Botanical Expeditions and Visual Culture in the Hispanic Enlightenment.* Chicago: University of Chicago Press, 2012.

Boke, Charis. "Regrounding Practice, Unsettling Knowledge: Plant Medicine in Settler Colonial Contexts." *Ethnobotanical Assembly,* Autumn 2020. www.tea-assembly. com.

Bolker, Jamie. "Stowe's Birds: Jim Crow and the Nature of Resistance in *Dred.*" *J19* 6, no. 2 (2018): 237–257.

Braun, Juliane. "Bioprospecting Breadfruit: Imperial Botany, Transoceanic Relations, and the Politics of Translation." *Early American Literature* 54, no. 3 (2019): 643–671.

Breck, Joseph. *The Flower-Garden; or, Breck's Book of Flowers.* Boston: J.P. Jewett, 1851.

Breckenridge, Carol, Sheldon Pollock, Homi K. Bhabha, and Dipesh Chakrabarty, eds. *Cosmopolitanism.* Durham, NC: Duke University Press, 2002.

Brickhouse, Anna. "Hawthorne in the Americas: Frances Calderón de la Barca, Octavio Paz, and the Mexican Genealogy of 'Rappaccini's Daughter.'" *PMLA* 113, no. 2 (March 1998): 227–242.

———. *Transamerican Literary Relations and the Nineteenth Century Public Sphere.* Cambridge: Cambridge University Press, 2007.

Brodhead, Richard. *Culture of Letters: Scenes of Reading and Writing in Nineteenth-Century America*. Chicago: University of Chicago Press, 1993.

Brontë, Charlotte. *Jane Eyre*. 1847. New York: Penguin, 2006.

Brown, Bill. *A Sense of Things: The Object Matter of American Literature*. Chicago: University of Chicago Press, 2003.

Brown, Gillian. *Domestic Individualism: Imagining Self in Nineteenth Century America*. Berkeley: University of California Press, 1992.

Browne, Janet. "Botany for Gentlemen: Erasmus Darwin and 'The Loves of the Plants.'" *Isis* 80, no. 4 (December 1989): 592–621.

———. *Charles Darwin: A Biography*. Vol. 2: *The Power of Place*. Princeton: Princeton University Press, 2003.

Buell, Lawrence. *The Environmental Imagination: Thoreau, Nature Writing, and the Formation of American Culture*. Cambridge, MA: Belknap, 1996.

Burr, Zofia. *Of Women, Poetry, and Power: Strategies of Address in Dickinson, Miles, Brooks, Lorde and Angelou*. Champaign: University of Illinois Press, 2002.

Byrd, Jodi. *The Transit of Empire: Indigenous Critiques of Colonialism*. Minneapolis: University of Minnesota Press, 2011.

Card, Kenton, dir. *Geographies of Racial Capitalism with Ruth Wilson Gilmore*. Antipode Foundation Film, 2020.

Carlyle, Thomas. "Occasional Discourse on the Negro Question." *Fraser's Magazine* 40 (December 1849).

Carney, Judith. *Black Rice: The African Origins of Rice Cultivation in the Americas*. Cambridge, MA: Harvard University Press, 2002.

Carney, Judith, and Richard Nicholas Rosomoff. *In the Shadow of Slavery: Africa's Botanical Legacy in the Atlantic World*. Berkeley: University of California Press, 2009.

Carr, Frederick Frances, Jr. "The Southern Planter, 1841–1861." Master's thesis, University of Richmond, 1971.

Carruthers, Clifford. "The 'Povera Picciola' and *The Scarlet Letter*." *Papers on Language and Literature* 7, no. 1 (1971): 90–94.

Casid, Jill. "Inhuming Empire: Islands as Colonial Nurseries and Graves." In *The Global Eighteenth Century*, edited by Felicity Nussbaum. Baltimore: Johns Hopkins University Press, 2003.

———. *Sowing Empire: Landscape and Colonization*. Minneapolis: University of Minnesota Press, 2005.

Chakrabarty, Dipesh. "The Climate of History: Four Theses." *Critical Inquiry* 35, no. 2 (Winter 2009): 197–222.

———. *Provincializing Europe: Postcolonial Thought and Historical Difference* (Princeton: Princeton University Press, 2007).

Child, John Lewis, ed. *Child's Fall Catalogue of Bulbs and Plants That Bloom*. New York: Child's Fall, 1893.

Child, Lydia Maria. *Evenings in New England: Intended for Juvenile Amusement and Instruction*. Boston: Cummings, Hilliard, 1824.

———. *Hobomok and Other Writings on Indians*. Edited by Carolyn Karcher. New Brunswick, NJ: Rutgers University Press, 1986.

———. *A Romance of the Republic*. Boston: Ticknor and Fields, 1867.

Chow, Juliana. "'Because I See—New Englandly—': Seeing Species in the Nineteenth Century and Emily Dickinson's Regional Specificity." *ESQ: A Journal of the American Renaissance* 60 (2014): 413–449.

———. *19th Century American Literature and the Discourse of Natural History*. Cambridge: Cambridge University Press, forthcoming.

Christian Recorder. "News from Dr. Livingston." October 11, 1862.

Claborn, John. *Civil Rights and the Environment in African-American Literature, 1895–1941*. London: Bloomsbury, 2018.

Coale, Samuel. "The Romance of Mesmerism." In *Studies in the American Renaissance*, edited by Joel Myerson, 271–288. Charlottesville: University of Virginia Press, 1994.

Cobbett, William. *The American Gardener*. New York: O. Judd, 1819.

Cohen, Benjamin. *Notes from the Ground: Science, Soil, and Society in the American Countryside*. New Haven, CT: Yale University Press, 2009.

Colacurcio, Michael. *The Province of Piety: Moral History in Hawthorne's Early Tales*. Cambridge, MA: Harvard University Press, 1984.

Conron, John. *American Picturesque*. University Park: Pennsylvania State Press, 2000.

Constitution of the New-York State Horticultural Society. New York, 1824. www.biodiversitylibrary.org.

Cooper, J. W. *The Experienced Botanist or Indian Physician, Being a New System of Practice Founded on Botany*. 1833. Lancaster, 1840.

Crevecoeur, J. Hector St. John de. *Letters from an American Farmer and Sketches of an Eighteenth-Century American*. Edited by Albert E. Stone. 1782. New York: Penguin, 1981.

Cronon, William. *Changes in the Land: Indians, Colonists, and the Ecology of New England*. 1983. New York: Hill & Wang, 2003.

Cronon, William, and Michael Pollan. "Out of the Wild: A Conversation between William Cronon and Michael Pollan." *Orion Magazine*, November/December 2013.

Crosby, Alfred. *The Columbian Exchange: Biological and Cultural Consequences of 1492*. Westport, CT: Greenwood, 1973.

———. *Ecological Imperialism: The Biological Expansion of Europe, 900–1900* Cambridge: Cambridge University Press, 2004.

Crumbley, Paul. "Dickinson's Correspondence and the Politics of Gift-Based Circulation." In *Reading Dickinson's Letters: Essays*, edited by Jane Donahue Eberwein and Cindy MacKenzie, 28–55. Amherst: University of Massachusetts Press, 2009.

Daily News. "Kew Gardens in Winter." 1886. Kewensia Collection, Kew Gardens, London.

D'Amore, Maura. "Thoreau's Unreal Estate: Playing House at Walden Pond." *New England Quarterly* 82, no. 1 (March 2009): 56–79.

Darwin, Charles. *On the Origin of Species*. 1859. New York: Appleton, 1860.

Darwin, Charles, and Francis Darwin. *The Power of Movement in Plants*. 1880. New York: Appleton, 1897.

Darwin, Erasmus. "The Loves of Plants." In *The Botanic Garden: A Poem, in Two Parts*. 1789. New York: T. & J. Swords, 1798.

———. *Phytologia: or, The Philosophy of Agriculture and Gardening*. Dublin: P. Byrne, 1800.

———. *Zoonomia: or, The Laws of Organic Life in Three Parts*. Vol. 1. Boston: Thomas and Andrews, 1803.

Daston, Lorraine, ed. *Things That Talk: Object Lessons from Art and Science*. Cambridge, MA: MIT Press, 2004.

Daunton, Martin J. "Introduction." In *The Organisation of Knowledge in Victorian Britain*, edited by Martin J. Daunton. Oxford: Oxford University Press, 2005.

Davidson, Cathy. *Revolution and the Word: The Rise of the American Novel*. Oxford: Oxford University Press, 1988.

Dearborn, H. A. S. "An Address Delivered before the Massachusetts Horticultural Society on the Celebration of Their First Anniversary." Boston: J.T. Buckingham, 1929.

———. "Letter to David Porter, Esq., Charge D'Affairs of the United States at the Ottoman Porte. Proceedings of the Massachusetts Horticultural Society, 18th of May, 1833." *New England Farmer, and Horticultural Register*, May 22, 1833.

———. "Mass Horticultural Society." *New England Farmer, and Horticultural Register*, May 22, 1833, 354.

———. "New Fruits and Ornamental Plants." *Horticultural Register, and Gardener's Magazine*, September 1, 1835, 324.

de Candolle, Augustin Pyramus. *Elements of the Philosophy of Plants, Containing the Principles of Scientific Botany*. Edited by Kurt Sprengel. 1821. Cambridge: Cambridge University Press, 2011.

DeLoughrey, Elizabeth. *Allegories of the Anthropocene*. Durham, NC: Duke University Press, 2019.

Dickinson, Emily. *Letters of Emily Dickinson*. Vol. 1. Edited by Mabel Loomis Todd. Boston: Roberts Brothers, 1894.

———. *The Poems of Emily Dickinson (Variorum Edition)*. Edited by R. W. Franklin. Cambridge, MA: Belknap, 1998.

———. *Selected Letters*. Edited by Thomas H. Johnson. Cambridge, MA: Belknap, 1971.

Dierks, Konstantin. *In My Power: Letter Writing and Communications in Early America*. Philadelphia: University of Pennsylvania Press, 2009.

Dillon, Elizabeth Maddock. "Religion and Geopolitics in the New World." *Early American Literature* 45, no. 1 (January 2010): 193–202.

———. "Sentimental Aesthetics." *American Literature* 76, no. 3 (September 2004): 495–523.

Dimock, Wai Chee. "Debasing Exchange: Edith Wharton's *The House of Mirth*." *PMLA* 100, no. 5 (October 1985): 783–791.

Doolan, Andy. "Blood, Republicanism, and the Return of George Washington: A Response to Shirley Samuels." *American Literary History* 20, nos. 1–2 (2007): 76–82.

Douglass, Frederick. "Address Delivered by Hon. Frederick Douglass, at the Third Annual Fair of the Tennessee Colored Agricultural and Mechanical Association." Washington, DC: New National Era and Citizen Print, 1873. Manuscript/Mixed Material in Frederick Douglass Papers, Library of Congress. www.loc.gov.

———. *Frederick Douglass: Selected Speeches and Writings*. Edited by Eric Foner. Chicago: Lawrence Hill Books, 1999.

———. "A Lecture on Our Nation's Capital." Exhibition Catalogue, Anacostia Neighborhood Museum, Smithsonian Institution. https://lccn.loc.gov.

———. "Letter to Harriet Beecher Stowe, March 8, 1853." In Douglass, *Frederick Douglass: Selected Speeches and Writings*.

———. *My Bondage and My Freedom*. 1855. New York: Penguin, 2003.

———. *Narrative of the Life of Frederick Douglass, an American Slave*. 1845. New York: Penguin, 2014.

———. "Pumpkins." *North Star*, October 19, 1849.

Downing, Andrew Jackson. *Treatise on the Theory and Practice of Landscape Gardening, Adapted to North America*. New York: Wily and Putnam, 1841.

Doyle, Laura. *Freedom's Empire: Race and the Rise of the Novel in Atlantic Modernity, 1640–1940*. Durham, NC: Duke University Press, 2008.

Drayton, Richard. *Nature's Government: Science, Imperial Government, and the "Improvement" of the World*. New Haven, CT: Yale University Press, 2000.

Dupuy, Edward. "Linguistic Mastery and the Garden of the Chattel in Frederick Douglass' 'Narrative.'" *Mississippi Quarterly* 44, no. 1 (1990–1991): 23–33.

Duquette, Elizabeth. "The Republican Mammy? Imagining Civic Engagement in *Dred*." *American Literature* 80, no. 1 (March 2008): 1–28.

Eberwein, Jane Donahue. "Outgrowing Genesis: Dickinson, Darwin, and the Higher Criticism." In *Emily Dickinson and Philosophy*, edited by Jed Deppman, Marianne Noble, and Gary Lee Stonum. Cambridge: Cambridge University Press, 2013.

Eclectic Magazine of Foreign Literature. "Pleasure of Botany and Gardening." January 1848.

Eclectic Magazine of Foreign Literature. "Can We Separate Animals from Plants." 90 (May 1878).

Edelman, Lee. *No Future: Queer Theory and the Death Drive*. Durham, NC: Duke University Press, 2004.

E.H.C. "Gleanings." *Horticulturalist and Journal of Rural Art and Rural Taste*, May 1863.

———. "Sensibility of Nature." *Ohio Farmer*, August 27, 1870.

Ellis, Cristin. "Amoral Abolitionism: Frederick Douglass and the Environmental Case against Slavery." *American Literature* 86, no. 2 (June 2014): 275–304.

Endersby, Jim. *Imperial Nature: Joseph Hooker and the Practices of Victorian Science*. Chicago: University of Chicago Press, 2008.

Ernest, John. *A Nation within a Nation: Organizing African-American Communities before the Civil War*. Lanham, MD: Ivan R. Dee, 2011.

Fagan, Benjamin. *The Black Newspaper and the Chosen Nation*. Athens: University of Georgia Press, 2016.

Family Magazine. "Botany." May 1836.

Farmer, Jared. *Trees in Paradise: A California History.* New York: Norton, 2013.

Farooq, Nihad. *Undisciplined: Science, Ethnography and Personhood in the Americas, 1830–1940.* New York: New York University Press, 2016.

Farr, Judith. *The Gardens of Emily Dickinson.* Cambridge, MA: Harvard University Press, 2005.

Favini, John. "What if Competition Isn't as 'Natural' as We Think?" *Slate,* January 23, 2020.

Favret, Mary. *War at a Distance: Romanticism and the Making of Modern Wartime.* Princeton: Princeton University Press, 2009.

F.D. "The Smith Lands." *North Star,* January 5, 1849.

Feeley, Lynne. "Plants and the Problem of Authority in the Antebellum United States." In Laist, *Plants and Literature,* 53–74.

———. "Swamps, Squash, Slavery." *Avidly,* May 30, 2013.

Feerick, Jean. *Strangers in Blood: Relocated Race in the Renaissance.* Toronto: University of Toronto Press, 2010.

Fett, Sharla. *Working Cures: Healing, Health, and Power on Southern Slave Plantations.* Chapel Hill: University of North Carolina Press, 2002.

Finlay, Ian Hamilton. "Detached Sentences on Gardening." In *Selections.* Berkeley: University of California Press, 2012.

Finseth, Ian Frederick. *Shades of Green: Visions of Nature in the Literature of American Slavery, 1770–1860.* Athens: University of Georgia Press, 2007.

Fliegelman, Jay. "Introduction." In *Wieland and Memoirs of Carwin the Biloquist.* New York: Penguin, 1991.

Folsom, Ed, and Kenneth Price. "Dickinson, Slavery, and the San Domingo Movement." Whitman Archive, n.d. http://whitmanarchive.org.

Foner, Eric. *Reconstruction: America's Unfinished Revolution, 1863–1877.* New York: Harper, 2014.

Foreman, P. Gabrielle. "Sentimental Abolition in Douglass' Decade: Revision, Erotic Conversion, and the Politics of Witnessing in 'The Heroic Slave' and *My Bondage and My Freedom.*" In *Criticism and the Color Line: Desegregating American Literary Studies,* edited by Henry B. Wonham, 191–204. New Brunswick, NJ: Rutgers University Press, 1996.

Fraiman, Susan. *Extreme Domesticity: A View from the Margins.* New York: Columbia University Press, 2017.

Frederick Douglass' Paper. "The Colored People's Agricultural Fair at Geneva." October 13, 1854.

———. "He Who Is a Good Theoretical Agriculturalist. . . ." June 8, 1855.

———. "Make Your Sons Mechanics and Farmers—Not Waiters Porters and Barbers." March 18, 1853.

———. "The Word 'White.'" March 17, 1854.

Freedgood, Elaine. *The Ideas in Things: Fugitive Meaning in the Victorian Novel.* Chicago: University of Chicago Press, 2006.

Fretwell, Erica. "Emily Dickinson in Domingo." *J19* 1, no. 1 (Spring 2013): 71–96.

Fuller, Margaret. "Autobiographical Romance." In *The Essential Margaret Fuller*, edited by Jeffrey Steele. New Brunswick, NJ: Rutgers University Press, 1992.

———. *Summer on the Lakes*. In *The Essential Margaret Fuller*, edited by Jeffrey Steele. New Brunswick, NJ: Rutgers University Press, 1992.

Fullilove, Courtney. *The Profit of the Earth: The Global Seeds of American Agriculture*. Chicago: University of Chicago Press, 2017.

Gagliano, Monica, John C. Ryan, and Patricia Vieira, eds. *The Language of Plants*. Minneapolis: University of Minnesota Press, 2017.

Gardner, Eric. *Black Print Unbound: The Christian Recorder, African American Literature, and Periodical Culture*. Oxford: Oxford University Press, 2015.

Gaudio, Michael. "The Elements of Botanical Art: William Bartram, Benjamin Smith Barton, and the Scientific Imagination." In *William Bartram: The Search for Nature's Design*, edited by Thomas Hillock and Nancy E. Hoffman. Athens: University of Georgia Press, 2010.

Genesee Farmer and Gardener's Journal. "The Action of Poison on Vegetable Structure—Rotation of Crops." April 12, 1834.

George, Samantha. *Botany, Sexuality and Women's Writing: From Modest Shoot to Forward Plant*. New York: Manchester University Press, 2007.

Gerhardt, Christine. "'Often Seen—but Seldom Felt': Emily Dickinson's Reluctant Ecology of Place." *Emily Dickinson Journal* 15, no. 1 (2006): 56–78.

———. *A Place for Humility: Whitman, Dickinson, and the Natural World*. Iowa City: University of Iowa Press, 2014.

Gianquitto, Tina. *"Good Observers of Nature": American Women and the Study of the Natural World 1820–1885*. Athens: University of Georgia Press, 2007.

Giles, Paul. "'The Earth Reversed Her Hemispheres': Dickinson's Global Antipodality." *Emily Dickinson Journal* 20, no. 1 (2011): 1–21.

Glave, Dianne D. and Mark Stoll, eds., *"To Love the Wind and the Rain": African Americans and Environmental History*. Pittsburgh: University of Pittsburgh Press, 2005.

Godey's Lady's Book. "Picciola." 35, no. 7 (1847): 57.

———. "The Principle of Life." February 1933.

Goldstein, Amanda Jo. *Sweet Science: Romantic Materialism and the New Logics of Life*. Chicago: University of Chicago Press, 2017.

Goudie, Sean. *Creole America: The West Indies and the Formation of Literature and Culture in the New Republic*. Philadelphia: University of Pennsylvania Press, 2006.

Gray, Asa. *The Botanical Text-Book: An Introduction to Scientific Botany*. 4th ed. New York: Putman, 1853.

Greeley, Horace. "Farm Life." *North Star*, October 20, 1848.

Greven, David. *Men Beyond Desire: Manhood, Sex, and American Literature*. New York: Palgrave Macmillan, 2012.

Guha, Ramachandra, and Joan Martinez-Alier. *Varieties of Environmentalism: Essays North and South*. New York: Routledge, 1997.

Guthrie, James R. "Darwinian Dickinson: The Scandalous Rise and Noble Fall of the Common Clover." *Emily Dickinson Journal* 16, no. 1 (2007): 73–91.

Hale, Sarah Josepha. *Flora's Interpreter; or, The American Book of Flowers and Sentiments*. Boston: Marsh, Capen and Lyon, 1832.

Hamner, Everett. "Here's to Unsuicide: An Interview with Richard Powers." *Los Angeles Review of Books*, April 7, 2018.

Hannickel, Erica. *Empire of Vines: Wine Culture in America*. Philadelphia: University of Pennsylvania Press, 2013.

———. "George Engelmann and 19th Century Cactus Collection and Circulation." Paper, C19, March 2018.

Haraway, Donna. *When Species Meet*. Minneapolis: University of Minnesota Press, 2008.

Haraway, Donna, Noboru Ishikawa, Scott Gilbert, Kenneth Olwig, Anna Tsing, and Nils Bubandt. "Anthropologists Are Talking about the Anthropocene." *Ethnos*, November 2015, 535–564.

Hartman, Saidiya. *Lose Your Mother: A Journey along the Atlantic Slave Route*. New York: Farrar, Straus and Giroux, 2008.

Haskell, David George. *The Forest Unseen: A Year's Watch in Nature*. New York: Penguin, 2013.

Hatton, Joseph. "A Day with Sir Joseph Hooker at Kew." *Harper's Monthly Magazine*, November 1884.

Hawthorne, Nathaniel. *American Notebooks*. Edited by Mitchell Simpson. 1868. Columbus: Ohio State University Press, 1972.

———. *The Blithedale Romance*. New York: Penguin, 1983.

———. *The House of the Seven Gables*. Edited by Milton Stern. 1851. New York: Penguin, 1981.

———. *Our Old Home, and English Note-Books*. 1863. Boston: Houghton Mifflin, 1912.

———. *Passages from the American Notebooks of Nathaniel Hawthorne*. Vol. 2. London: Smith, Elder, 1868.

———. *The Scarlet Letter*. Edited by Thomas Connolly. 1850. New York: Penguin, 2002.

———. *Selected Tales and Sketches*. Edited by Michael Colacurcio. New York: Penguin, 1987.

Hedrick, Joan. *Harriet Beecher Stowe: A Life*. Oxford: Oxford University Press, 1995.

Heise, Ursula K. *Sense of Place and Sense of Planet: The Environmental Imagination of the Global*. Oxford: Oxford University Press, 2008.

Helper, Hinton. *The Impending Crisis of the South and How to Meet It*. New York: Burdick Brothers, 1857.

Henchman, Anna. "Stone Deaf: Sentient Surfaces and Penetrability." Paper, International Conference on Narrative, MIT, March 27, 2014.

Hendler, Glenn. *Public Sentiments: Structures of Feeling in Nineteenth-Century American Literature*. Chapel Hill: University of North Carolina Press, 2001.

Hertz, W. Watford. Letter from W. Watford Hertz to Sir William Hooker, December 31, 1850. Kewensia Collection, Kew Gardens, London.

Higginson, Thomas Wentworth. "A Step Backward?" *Chap-book: Semi-Monthly*, November 1, 1894, 330D.

Hindi, Noor. "Fuck Your Lecture on Craft, My People Are Dying." *Poetry*, December 2020. www.poetryfoundation.org.

Historical Note. "Finding Aid for Emily Dickinson Botanical Specimens, Undated." Houghton Library.

Hitchcock, Edward. "Edward Hitchcock Classroom Lecture Notes, 'Botany.'" N.d. [1826–1855]. Amherst College Archives and Special Collections, box 10, folder 2.

Holbrook, Josiah. "Democracy of Science—No. 22." *National Era*, February 17, 1853.

Holmes, Richard. *The Age of Wonder: How the Romantic Generation Discovered the Beauty and Terror of Science*. New York: Vintage, 2008.

Horticulturalist and Journal of Rural Art and Rural Taste. "The Horticultural Festival at Faneuil Hall, Boston." November 1848.

———. "The New Water Lily—Victoria Regia." December 1850.

Houle, Karen L. F. "Animal, Vegetable, Mineral: Ethics as Extension or Becoming? The Case of Becoming-Plant." *Journal for Critical Animal Studies* 9, no. 1/2 (2011).

Howard, June. "What Is Sentimentality?" *American Literary History* 11, no. 1 (1999): 63–81.

Howell, William Huntting. "In the Realms of Sensibility." *American Literary History* 25, no. 2 (Summer 2013): 406–417.

Humboldt, Alexander von. *Views of Nature*. Edited by Laura Dassow Walls and Stephen T. Jackson. Chicago: University of Chicago Press, 2014.

Iannini, Christopher. *Fatal Revolutions: Natural History, West Indian Slavery, and the Routes of American Literature*. Chapel Hill: University of North Carolina Press, 2012.

Jabr, Ferris. "The Social Life of Forests." *New York Times*, December 2, 2020.

Jackson, Cassandra. *Barriers between Us: Interracial Sex in Nineteenth-Century American Literature*. Bloomington: Indiana University Press, 2004.

Jackson, Holly. "The Transformation of American Property in *The House of the Seven Gables*." *ESQ* 56, no. 3 (2010): 269–292.

Jackson-Houlston, Caroline. "'Queen Lilies'? The Interpenetration of Scientific, Religious and Gender Discourses in Victorian Representations of Plants." *Journal of Victorian Culture* 11, no. 1 (2006).

Jacobs, Harriet. *Incidents in the Life of a Slave Girl*. Boston, 1861.

Jacoby, Karl. *Crimes Against Nature: Squatters, Poachers, Thieves and the Hidden History of American Conservation*. Oakland: University of California Press, 2014.

Jafari, Ghazal, Pierre Bélanger, and Pablo Escudero. *A Botany of Violence: Across 529 Years of Resistance and Resurgence*. Novato, CA: Goff Books, 2021.

Jager, Colin. *The Book of God: Secularization and Design in the Romantic Era*. Philadelphia: University of Pennsylvania Press, 2006.

Jahren, Hope. *Lab Girl*. New York: Knopf, 2016.

James, Henry. *Hawthorne*. 1879. New York: Harper Brothers, 1899.

James, Jennifer. "'Buried in Guano': Race, Labor, and Sustainability." *ALH* 24, no. 1 (2012): 115–142.

Jefferson, Thomas. *Notes on the State of Virginia*. Edited by Frank Shuffelton. 1785. New York: Penguin, 1998.

Johnson, Myles E. "About." Underground Plant Trade, n.d. www.freedemplants.com.

Johnson, Samuel. "Rasselas: A Tale." In *The Works of Samuel Johnson, LL.D.* Vol. 1. 1759. London: W. Pickering, 1825.

Kades, Eric. "History and Interpretation of the Great Case of *Johnson v. M'Intosh*." *Law and History Review* 19, no. 1 (2001): 67–116.

Kaplan, Amy. *The Anarchy of Empire in the Making of US Culture*. Cambridge, MA: Harvard University Press, 2002.

Karafilis, Maria. "Spaces of Democracy in Harriet Beecher Stowe's *Dred*." *Arizona Quarterly* 55, no. 3 (1999): 23–49.

Karcher, Carolyn. *The First Woman in the Republic: A Cultural Biography of Lydia Maria Child*. Durham, NC: Duke University Press, 1994.

Keeney, Elizabeth. *The Botanizers: Amateur Scientists in Nineteenth-Century America*. Chapel Hill: University of North Carolina Press, 1992.

Kelley, Theresa M. *Clandestine Marriage: Botany and Romantic Culture*. Baltimore: Johns Hopkins University Press, 2012.

Kenrick, William. *Farmer's Register: A Monthly Publication* 7, no. 4 (April 30, 1839).

Kerber, Linda K. "The Abolitionist Perception of the Indian." *Journal of American History* 62, no. 2 (September 1975).

Kilcup, Karen. *Fallen Forests: Emotion, Embodiment, and Ethics in American Women's Environmental Writing, 1781–1924*. Athens: University of Georgia Press, 2013.

———. "Feeling American in the Poetic Republic." *Nineteenth-Century Literature* 70, no. 3 (2015): 299–335.

Kimmerer, Robin Wall. *Braiding Sweetgrass: Indigenous Wisdom, Scientific Knowledge and the Teachings of Plants*. Minneapolis: Milkweed, 2013.

King, Amy. *Bloom: The Botanical Vernacular in the English Novel*. Oxford: Oxford University Press, 2003.

Kingbury, Noel. *Hybrid: The History and Practice of Plant Breeding*. Chicago: University of Chicago Press, 2009.

Kirksey, Eben. *Emergent Ecologies*. Durham, NC: Duke University Press, 2015.

Kirksey, Eben, Craig Schuetze, and Stefan Helmreich, eds. *The Multispecies Salon*. Durham, NC: Duke University Press, 2014.

Kohn, Eduardo. *How Forests Think: Towards an Anthropology beyond the Human*. Berkeley: University of California Press, 2013.

Kolodny, Annette. *The Lay of the Land: Metaphor as Experience and History in American Life and Letters*. Chapel Hill: University of North Carolina Press, 1975.

Korobkin, Laura. "The Scarlet Letter of the Law: Hawthorne and Criminal Justice." *NOVEL* 30, no. 2 (1997): 193–217.

Kot, Paula. "Engendering Identity: Doubts and Doubles in Lydia Maria Child." *Hypatia* 19, no. 2 (Spring 2004).

Kupperman, Karen. "Fear of Hot Climates in the Anglo-American Colonial Experience." *William and Mary Quarterly* 41, no. 2 (April 1984): 213–240.

Kutzinski, Vera. "Borders and Bodies: The United States, America, and the Caribbean." *New Centennial Review* 1, no. 2 (Fall 2001): 55–88.

Ladies' Companion. "Picciola." 10 (January 1, 1839): 147.

LaFleur, Greta. *The Natural History of Sexuality in Early America.* Baltimore: Johns Hopkins University Press, 2018.

Laist, Randy, ed. *Plants and Literature: Essays in Critical Plant Studies.* Boston: Brill/ Rodolpi, 2013.

Lambert, Lilly. *The American Forest: or, Uncle Philip's Conversations with the Children about the Trees of America.* New York: Harper, 1834.

Lane, Lea. "A Lovely Bunch of Coconuts." *Winterthur Unreserved*, November 25, 2015. http://museumblog.winterthur.org.

Latour, Bruno. *We Have Never Been Modern.* 1991. Translated by Catherine Porter. Cambridge, MA: Harvard University Press, 1993.

Lee, Maurice S. *Uncertain Chances: Science, Skepticism, and Belief in Nineteenth-Century American Literature.* Oxford: Oxford University Press, 2012.

LeMenager, Stephanie. *Living Oil: Petroleum Culture in the American Century.* Oxford: Oxford University Press, 2014.

———. "Love and Theft; or, Provincializing the Anthropocene." *PMLA* 136, no. 1 (2021): 102–109.

———. *Manifest and Other Destinies: Territorial Fictions of the Nineteenth-Century United States.* Lincoln: University of Nebraska Press, 2004.

Lemire, Elise. *"Miscegenation": Making Race in America.* Philadelphia: University of Pennsylvania Press, 2002.

Levine, George. *Dying to Know: Scientific Epistemology and Narrative in Victorian England.* Chicago: University of Chicago Press, 2002.

Levine, Robert S. *Dislocating Race and Nation: Episodes in Nineteenth-Century American Literary Nationalism.* Chapel Hill: University of North Carolina Press, 2008.

Lewis, Andrew J. *A Democracy of Facts: Natural History in the Early Republic.* Philadelphia: University of Pennsylvania Press, 2010.

———. "Gathering for the Republic: Botany in Early Republic America." In *Colonial Botany: Science, Commerce, and Politics in the Early Modern World*, edited by Londa Schiebinger and Claudia Swan, 69. Philadelphia: University of Pennsylvania Press, 2005.

Lindley, John. "Remarks on Hybridising Plants." Reprinted in *Horticulturalist and Journal of Rural Art and Rural Taste*, September 1847.

———. *The Theory of Horticulture; or, An Attempt to Explain the Principle Operations of Gardening upon Physiological Principles.* London: Longman, Orme, Brown, Green, and Longmans, 1840.

Linnaeus, Carl. *Philosphia Botanica.* 1751. Translated by Stephen Freer. Oxford: Oxford University Press, 2003.

Literary World. "How Plants Behave." January 1, 1873.

Loman, Andrew. "'More Than a Three-Pence': Crises of Value in Hawthorne's 'My Kinsman, Major Molineux.'" *PMLA* 126, no. 2 (2011): 345–362.

Looby, Christopher. "Flowers of Manhood: Race, Sex and Floriculture from Thomas Wentworth Higginson to Robert Mapplethorpe." *Criticism* 37, no. 1 (Winter 1995): 109–156.

Loughran, Trish. *The Republic in Print: Print Culture in the Age of U.S. Nation Building, 1770–1870.* New York: Columbia University Press, 2009.

Lowe, Lisa. *The Intimacies of Four Continents.* Durham, NC: Duke University Press, 2015.

Luedke, Luther. *Nathaniel Hawthorne and the Romance of the Orient.* Bloomington: Indiana University Press, 1989.

Lynch, Deirdre. "'Young Ladies Are Delicate Plants': Jane Austen and Greenhouse Romanticism." *ELH* 77, no. 3 (2010): 689–729.

Malm, Andreas. "In Wildness Is the Liberation of the World: On Maroon Ecology and Partisan Nature," *Historical Materialism* 26, no. 3 (2018): 3–37.

Mann, Charles. *1491: New Revelations of the Americas Before Columbus.* New York: Vintage, 2006.

Mann, Horace. *Lectures on Education.* Boston: Fowle & Capen, 1845.

Manning, Robert. *History of the Massachusetts Horticultural Society 1829–1878.* Boston: Printed for the Society, 1880.

Marder, Michael. *The Philosopher's Plant: An Intellectual Herbarium.* New York: Columbia University Press, 2014.

———. *Plant-Thinking: A Philosophy of Vegetal Life.* New York: Columbia University Press, 2013.

Marrs, Cody. *Nineteenth-Century American Literature and the Long Civil War.* Cambridge: Cambridge University Press, 2015.

Martin, Aryn, Natasha Myers, and Ana Viseu. "The Politics of Care in Technoscience, Introduction to a Special Issue of Social Studies of Science." *Social Studies of Science* 45, no. 5 (September 28, 2015): 625–641.

Marvel, Andrew. "The Mower against Gardens." In *Complete Poems.* New York: Penguin, 1996.

Marx, Leo. *The Machine in the Garden: Technology and the Pastoral Ideal in America.* Oxford: Oxford University Press, 1964.

McClarin, Ka'mal. "Cedar Hill: Frederick Douglass's Rustic Sanctuary." National Park Service, October 15, 2021. www.nps.gov.

McCune Smith, James. 1855. "Introduction." In Douglass, *My Bondage and My Freedom.*

McDowell, Marta. *Emily Dickinson's Gardens.* New York: McGraw-Hill, 2004.

McFeely, William S. *Frederick Douglass.* New York: Norton, 1991.

Mc Vickar. "On Education." *Cultivator,* August 1836.

Medoro, Dana. "'Looking into the Inmost Nature': Speculum and Sexual Selection in 'Rappaccini's Daughter.'" *Nathaniel Hawthorne Review* 35, no. 1 (Spring 2009): 70–86.

Meeker, Natania, and Antónia Szabari. *Radical Botany: Plants and Speculative Fiction.* New York: Fordham University Press, 2019.

Merish, Lori. *Sentimental Materialism: Gender, Commodity Culture and Nineteenth-Century American Literature.* Durham, NC: Duke University Press, 2000.

Michell, Roger, director. *Notting Hill.* London: PolyGram Filmed Entertainment, Working Title Films, 1999.

Mielke, Laura. *Moving Encounters: Sympathy and the Indian Question in Antebellum Literature.* Amherst: University of Massachusetts Press, 2008.

———. "Sentiment and Space in Lydia Maria Child's Native American Writings, 1824–1870." *Legacy* 21, no. 2 (2004): 172–192.

Milder, Robert. "Hawthorne and the Problem of New England." *American Literary History* 21, no. 3 (September 2009): 464–491.

Mill, John Stuart. *The Collected Works of John Stuart Mill.* Vol. 8. Edited by John M. Robson. Toronto: University of Toronto Press, 1974.

Miller, Cristanne. *Reading in Time: Emily Dickinson in the Nineteenth-Century.* Amherst: University of Massachusetts Press, 2012.

Miller, Daegan. "Reading Trees in Nature's Nation: Toward a Field Guide to Sylvan Literacy." *American Historical Review* 121, no. 4 (October 2016): 1114–1140.

———. *This Radical Land: A Natural History of American Dissent.* Chicago: University of Chicago Press, 2018.

Miller, Perry. *Errand into the Wilderness.* Cambridge, MA: Belknap, 1956.

Mitchell, Domnall. *Emily Dickinson: Monarch of Perception.* Amherst: University of Massachusetts Press, 2000.

Mitchell, Timothy. *Rule of Experts: Egypt, Technopolitics, Modernity.* Berkeley: University of California Press, 2002.

Morrison, Toni. "Home." In *The House That Race Built,* edited by Wahneema Lubiano. New York: Vintage, 1998.

Morton, Timothy. *The Ecological Thought.* Cambridge, MA: Harvard University Press, 2010.

Myers, Jeffrey. *Converging Stories: Race, Ecology, and Environmental Justice in American Literature.* Athens: University of Georgia Press, 2005.

Myers, Natasha. "Conversations on Plant Sensing: Notes from the Field." *NatureCulture* 3 (2015): 35–66.

National Magazine: Devoted to Literature, Art and Religion. "The Circassian Tribes and Schamyl." December 1854.

Nealon, Jeffrey. *Plant Theory: Biopower and Vegetable Life.* Palo Alto, CA: Stanford University Press, 2015.

Nelson, Dana D. *The Word in Black and White: "Race" in American Literature 1638–1867.* Oxford: Oxford University Press, 1994.

Nemerov, Alexander. *Still Life and Selfhood: The Body of Raphael Peele.* Berkeley: University of California Press, 2001.

New England Farmer. "The Concord Grape." 6, no. 4 (April 1854): 161.

New England Farmer, and Horticultural Register. "Grafting Fruit Trees." March 7, 1828.

New-Yorker. "Picciola." 3, no. 5 (April 22, 1837): 67.

New York Times. "A Flower Market." June 22, 1855.

Noble, Marianne. *The Masochistic Pleasures of Sentimental Literature*. Princeton: Princeton University Press, 2000.

North American Review. "Botany of the United States." 13, no. 32 (July 1821): 100–134.

Northup, Solomon. 1853. *Twelve Years a Slave*. New York: Norton, 2016.

Nuttall, Thomas. *The Genera of North American Plants, and a Catalogue of the Species to the Year 1817*. Vol. 1. Philadelphia: D. Heartt, 1818.

Ogden, Emily. *Credulity: A Cultural History of U.S. Mesmerism*. Chicago: University of Chicago Press, 2018.

Olmstead, Alan L., and Paul Webb Rhode. *Creating Abundance: Biological Innovation and American Agricultural Development*. Cambridge: Cambridge University Press, 2008.

Olmsted, Frederick Law. *Journey in the Seaboard Slave States*. New York: Dix and Edwards, 1856.

———. "Letters on the Productions, Industry and Resources of the Southern States." *New York Daily Times*, April 8, 1853. In *The Papers of Frederick Law Olmsted: Slavery and the South*, vol. 2, edited by Charles E. Beveridge and Charles Capen McLaughlin. Baltimore: Johns Hopkins University Press, 1981.

Osseo-Asare, Abena Dove. *Bitter Roots: The Search for Healing Plants in Africa*. Chicago: University of Chicago Press, 2014.

Otter, Samuel. "Stowe and Race." In *The Cambridge Companion to Harriet Beecher Stowe*, edited by Cindy Weinstein. Cambridge: Cambridge University Press, 2004.

Outka, Paul. *Race and Nature from Transcendentalism to the Harlem Renaissance*. New York: Palgrave Macmillan, 2008.

Pacheco, Derek. "'Vanished Scenes . . . Pictured in the Air': Hawthorne, Indian Removal and *The Whole History of Grandfather's Chair*." *Nathaniel Hawthorne Review* 36, no. 1 (Spring 2010): 186–212.

Page, Judith W., and Elise L. Smith. *Women, Literature, and the Domesticated Landscape: England's Disciples of Flora, 1780–1870*. Cambridge: Cambridge University Press, 2011.

Parish, Susan Scott. *American Curiosity: Cultures of Natural History in the Colonial British Atlantic World*. Chapel Hill: Omohundro Institute and University of North Carolina Press, 2006.

———. *The Flood Year 1927: A Cultural History*. Princeton: Princeton University Press, 2017.

Pauly, Philip. *Fruits and Plains: The Horticultural Transformation of America*. Cambridge, MA: Harvard University Press, 2008.

Pawley, Emily. "Accounting with the Fields: Chemistry and Value in Nutriment in American Agricultural Improvement, 1835–1860." *Science as Culture* 19, no. 4 (2010): 461–482.

Peel, Robin. *Emily Dickinson and the Hill of Science*. Madison, NJ: Fairleigh Dickinson University Press, 2010.

Perry, Tony. "In Bondage When Cold Was King: The Frigid Terrain of Slavery in Antebellum Maryland," *Slavery and Abolition* 38, no. 1 (2017): 23–36.

Petrino, Elizabeth. *Emily Dickinson and Her Contemporaries: Women's Verse in America, 1820–1885*. Hanover, NH: University Press of New England, 1998.

Phelps, Almira Hart Lincoln. *Botany for Beginners*. 1833. New York: F.J. Huntington, 1837.

———. *Familiar Lectures on Botany*. Hartford, CT: H. and F. Huntington, 1829.

———. *Familiar Lectures on Botany*. 1829. New York: Huntington and Mason Bros, 1854.

Phelps, Elizabeth Stuart. *The Story of Avis*. Boston: James R. Osgood, 1877.

Phillips, Dana. *The Truth of Ecology: Nature, Culture, and Literature in America*. Oxford: Oxford University Press, 2003.

Philo Florist. "On the Cultivation of Flowers." *Farmer's Cabinet* 1 (1838): 125.

Plotz, John. "Can the Sofa Speak? A Look at Thing Theory." *Criticism* 47, no. 1 (Winter 2005): 109–118.

Poe, Edgar Allan. "The Domain of Arnheim." In *Poetry, Tales, and Selected Essays*. 1846. New York: Library of America, 1996.

Pollan, Michael. "The Intelligent Plant: Scientists Debate a New Way of Understanding Flora." *New Yorker*, December 23, 2013.

———. "Weeds Are Us." *New York Times Magazine*, November 5, 1989.

Posmentier, Sonya. *Cultivation and Catastrophe: The Lyric Ecology of Modern Black Literature*. New York: New York University Press, 2017.

Powers, Richard. *The Overstory*. New York: Heinemann, 2018.

Pratt, Mary Louise. *Imperial Eyes: Travel Writing and Transculturation*. New York: Routledge, 1992.

Provincial Freeman. "Baby Show in Ohio." October 28, 1854.

Pruitt, Beth. "Reordering the Landscape: Science, Nature, and Spirituality at Wye House." PhD diss., University of Maryland at College Park, 2015.

Punch. "A Black Statue to Thomas Carlyle." Reprinted in *North Star*, February 22, 1850.

Purdy, Jedidiah. *After Nature: A Politics of the Anthropocene*. Cambridge, MA: Harvard University Press, 2015.

Putnam's Monthly Magazine. "A Chat about Plants." 3, no. 16 (April 1854).

Quarles, Benjamin. *Black Abolitionists*. Oxford: Oxford University Press, 1969.

Ray, Sarah Jaquette. "Risking Bodies in the Wild: The 'Corporeal Unconscious' of American Adventure Culture." In *Disability Studies and the Environmental Humanities: Towards an Eco-Crip Theory*, edited by Sarah Jaquette Ray and Jay Sibara. Lincoln: University of Nebraska Press, 2017.

Revkin, Andrew C. "Ecuador Constitution Grants Rights to Nature." *New York Times*, September 29, 2008.

Rezek, Joseph. *London and the Making of Provincial Literature: Aesthetics and the Transatlantic Book Trade*. Philadelphia: University of Pennsylvania Press, 2015.

Richards, Eliza. "'How News Must Feel When Traveling': Dickinson and Civil War Media." In *A Companion to Emily Dickinson*, edited by Martha Nell Smith and Mary Loeffelholz. Oxford: Blackwell, 2007.

Rifkin, Mark. *Settler Common Sense: Queerness and Everyday Colonialism in the American Renaissance*. Minneapolis: University of Minnesota Press, 2014.

Riskin, Jessica. *Science in the Age of Sensibility: The Sentimental Empiricists of the French Enlightenment*. Chicago: University of Chicago Press, 2002.

Riss, Arthur. "Sentimental Douglass." In *Cambridge Companion to Frederick Douglass*, edited by Maurice Lee. Cambridge: Cambridge University Press, 2009.

Ritvo, Harriet. *The Animal Estate: The English and Other Creatures in the Victorian Age*. Cambridge, MA: Harvard University Press, 1987.

———. "At the Edge of the Garden: Nature and Domestication in Eighteenth- and Nineteenth-Century Britain." *Huntington Library Quarterly* 55, no. 3 (Summer 1992): 363–378.

———. *Noble Cows and Hybrid Zebras: Essays on Animals and History*. Charlottesville: University of Virginia Press, 2010.

———. *The Platypus and the Mermaid: And Other Figments of the Victorian Classifying Imagination*. Cambridge, MA: Harvard University Press, 1998.

Robert Merry's Museum. "Wonderful Trees, No. 3—The India Rubber Tree." July 1, 1849.

Rogers, Elizabeth Barlow. *Writing the Garden: A Literary Conversation across Two Centuries*. Jaffrey, NH: David R. Godine, 2011.

Romero, Lora. *Home Fronts: Domesticity and Its Critics in the Antebellum United States*. Durham, NC: Duke University Press, 1997.

Ron, Ariel. "Summoning the State: Northern Farmers and the Transformation of American Politics in the Mid-Nineteenth Century." *Journal of American History* 103, no. 2 (September 2016): 347–374.

Rosenthal, Caitlin. "Slavery's Scientific Management: Masters and Managers." In *Slavery's Capitalism: A New History of American Economic Development*, edited by Sven Beckert and Seth Rockman, 62–86. Philadelphia: University of Pennsylvania Press, 2016.

Rosenthal, Debra. "Floral Counterdiscourse: Miscegenation, Ecofeminism, and Hybridity in Lydia Maria Child's *A Romance of the Republic*." *Women's Studies* 31, no. 2 (2002): 221–245.

Ruffin, Kimberly. *Black on Earth: African American Ecoliterary Traditions*. Athens: University of Georgia Press, 2010.

Rusert, Britt. *Fugitive Science: Empiricism and Freedom in Early African American Culture*. New York: New York University Press, 2017.

Sachs, Oliver. "The Mental Life of Plants and Worms, among Others." *New York Review of Books*, April 24, 2014.

Sagan, Lynn. "On the Origin of Mitosing Cells." *Journal of Theoretical Biology* 14, no. 3 (March 1967): 225–274.

Saintine, X. B. *Picciola: or, Captivity Captive*. 1836. Philadelphia: Carey, Lea & Blanchard, 1838.

Samuels, Shirley. "Women, Blood, and Contract." *ALH* 20, nos. 1–2 (2008): 57–75.

Sanborn, Geoffrey. "Introduction to the Nonhuman Turn Forum." *J19* 1 (Fall 2013).

Sandilands, Catriona. "Fear of a Queer Plant." *GLQ* 23, no. 3 (2017): 419–429.

Scaramelli, Caterina. *How to Make a Wetland: Water and Moral Ecology in Turkey.* Stanford, CA: Stanford University Press, 2021.

Schiebinger, Londa. *Plants and Empire: Colonial Bioprospecting in the Atlantic World.* Cambridge, MA: Harvard University Press, 2004.

Schoolman, Martha. *Abolitionist Geographies.* Minneapolis: University of Minnesota Press, 2014.

———. "Violent Places: Three Years in Europe and the Question of William Wells Brown's Cosmopolitanism." *ESQ* 58, no. 1 (2012): iv–35.

Schuller, Kyla. *The Biopolitics of Feeling: Race, Sex and Science in the Nineteenth Century.* Durham, NC: Duke University Press, 2018.

Scribner's Monthly. "Curiosities of Plant Life." 3, no. 6 (April 1872).

Seaton, Beverly. *The Language of Flowers: A History.* Charlottesville: University of Virginia Press, 1995.

Shakespeare, William. *The Winter's Tale.* In *The Arden Shakespeare Complete Works,* edited by Richard Proudfoot, Ann Thompson, and David Scott Kastan. 1623. London: Methuen Drama, 2011.

Shteir, Ann. *Cultivating Women, Cultivating Science: Flora's Daughters and Botany in England, 1760–1860.* Baltimore: Johns Hopkins University Press, 1994.

Sigourney, Lydia Howard. *Letters to Young Ladies.* Hartford, CT: P. Canfield, 1833.

———. *The Voice of Flowers.* 4th ed. Hartford, CT: H.S. Parsons, 1847.

Simard, Suzanne. *Finding the Mother Tree: Discovering the Wisdom of the Forest.* New York: Penguin, 2021.

Slotkin, Richard. *The Fatal Environment: The Myth of the Frontier in the Age of Industrialization, 1800–1890.* Norman: University of Oklahoma Press, 1998.

———. *Regeneration through Violence.* Norman: University of Oklahoma Press, 1973.

Smith, Henry Nash. *Virgin Land: The American West as Symbol and Myth.* Cambridge, MA: Harvard University Press, 1970.

Smith, Kimberly K. *African American Environmental Thought: Foundations.* Lawrence: University Press of Kansas, 2007.

Solnit, Rebecca. *River of Shadows.* New York: Penguin, 2004.

South Side. "For the Southern Planter: Superiority of the Farming Profession." *Southern Planter* 14, no. 7 (July 1852): 204.

Spary, Emma. *Utopia's Garden: French Natural History from Old Regime to Revolution.* Chicago: University of Chicago Press, 2000.

Spires, Derrick. *The Practice of Citizenship: Black Politics and Print Culture in the Early United States.* Philadelphia, University of Pennsylvania Press, 2019.

State v. Mann, 13 NC 263 (1829).

Stein, Rachel. *Shifting the Ground: American Women Writers' Revisions of Nature, Gender, and Race.* Charlottesville: University of Virginia Press, 1997.

Stewart, Mart. *"What Nature Suffers to Groe": Life, Labor, and Landscape on the Georgia Coast, 1680–1920.* Athens: University of Georgia Press, 2002.

Stoler, Ann. *Along the Archival Grain: Epistemic Anxieties and Colonial Common Sense.* Princeton: Princeton University Press, 2010.

Stone, Christopher. 1973. *Should Trees Have Standing? Law, Morality and the Environment*. 3rd ed. Oxford: Oxford University Press, 2010.

Stowe, Harriet Beecher. *Dred: A Tale of the Dismal Swamp*. Edited by Robert Levine. 1856. Chapel Hill: University of North Carolina Press, 2006.

———. "Harriet Beecher Stowe to Frederick Law Olmsted." Letter, June 1856. Houghton Library.

———. "House and Home Papers." *Atlantic Monthly* 14, no. 85 (November 1864).

———. "Introduction." In *A Library of Famous Fiction Embracing the Nine Standard Masterpieces of Imaginative Literature (Unabridged)*. New York: J.B. Ford, 1873.

———. "Meditations from Our Garden Seat." *Independent*, August 1855, 349.

———. *Oldtown Folks*. 1869. New Brunswick, NJ: Rutgers University Press, 1987.

———. *Sunny Memories of Foreign Lands*. Vol. 2. Boston: Phillips, Sampson, 1854.

———. *Uncle Tom's Cabin*. 1852. New York: Signet, 1966.

Stowe, Susan Monroe. "Harriet Beecher Stowe as a Mother." *Youth's Companion*, May 4, 1899.

Sullivan, Sherry. "A Redder Shade of Pale: The Indianization of Heroes and Heroines in Nineteenth-Century American Fiction." *Journal of the Midwest Modern Language Association* 20, no. 1 (Spring 1987).

Sundquist, Eric. *To Wake the Nations: Race in the Making of American Literature*. Cambridge, MA: Harvard University Press, 1993.

Sutter, Paul. "The World with Us: The State of American Environmental History." *Journal of American History* 100, no. 1 (June 2013): 94–119.

Sweet, Nancy. "Dissent and the Daughter in *A New England Tale* and *Hobomok*." *Legacy* 22, no. 2 (2005): 107–125.

Tamarkin, Eliza. *Anglophilia: Deference, Devotion, and Antebellum America*. Chicago: University of Chicago Press, 2008.

Tawil, Ezra. *The Making of Racial Sentiment: Slavery and the Birth of the Frontier Romance*. Cambridge: Cambridge University Press, 2008.

Taylor, Astra. "Who Speaks for the Trees?" *The Baffler* 32 (September 2016).

Thomson, George. "Speech of George Thompson, Esq., M.P., On Free Trade with India." *London Mercury*. Reprinted in *North Star*, January 7, 1848.

Thoreau, Henry David. *A Week on the Concord and Merrimack Rivers*. 1849. New York: Penguin, 1998.

———. *"Wild Apples" and Other Natural History Essays*. Edited by William Rossi. Athens: University of Georgia Press, 2002.

———. *Wild Fruits: Thoreau's Rediscovered Last Manuscript*. Edited by Bradley P. Dean. New York: Norton 2001.

———. *The Writings of Henry David Thoreau: Journal*. Edited by Bradford Torrey. Boston: Houghton Mifflin, 1906.

Tingley, Stephanie. "'Blossom[s] of the Brain': Women's Culture and the Poetics of Emily Dickinson's Correspondence." In *Reading Dickinson's Letters: Essays*, edited by Jane Donahue Eberwein and Cindy MacKenzie, 56–79. Amherst: University of Massachusetts Press, 2009.

Tobin, Beth. *Colonizing Nature: The Tropics in British Art and Letters*. Philadelphia: University of Pennsylvania Press, 2005.

Tompkins, Jane. *Sensational Designs: The Cultural Work of American Fiction, 1790–1860*. Oxford: Oxford University Press, 1986.

Trachtenberg, Alan. *Shades of Hiawatha: Staging Indians, Making Americans, 1880–1930*. New York: Hill & Wang, 2004.

Tresch, Jonathan. *The Romantic Machine: Utopian Science and Technology after Napoleon*. Chicago: University of Chicago Press, 2012.

Tsing, Anna. *The Mushroom at the End of the World: On the Possibility of Life in Capitalist Ruins*. Princeton: Princeton University Press, 2015.

Unnamed newspaper. "The Palm House." Glasgow, 1849. Kewensia Collection, Kew Gardens, London.

Valencius, Conevery Bolton. *"The Health of the Country": How American Settlers Understood Themselves and Their Land*. New York: Basic Books, 2002.

Valenti, Patricia Dunlavy, ed. "Sophia Peabody Hawthorne's *American Notebooks*." *Studies in the American Renaissance* (1998): 115–185.

Waddill v. Martin, 38 NC 562 (1845).

Wagner-McCoy, Sarah. "Virgilian Chesnutt: Eclogues of Slavery and Georgics of Reconstruction in the *Conjure Tales*." *ELH* 80, no. 1 (Spring 2013): 199–220.

Wald, Priscilla. "Terms of Assimilation: Legislating Subjectivity in the Emerging Republic." In *Cultures of United States Imperialism*, edited by Amy Kaplan and Donald E. Pease. Durham, NC: Duke University Press, 1993.

Wallace, David Foster. "Consider the Lobster." *Gourmet*, August 2004.

Walls, Laura Dassow. *Henry David Thoreau: A Life*. Chicago: University of Chicago Press, 2017.

———. *The Passage to Cosmos: Alexander von Humboldt and the Shaping of America*. Chicago: University of Chicago Press, 2009.

———. *Seeing New Worlds: Henry David Thoreau and Nineteenth-Century Natural Science*. Madison: University of Wisconsin Press, 1995.

Wampole, Christy. *Rootedness: The Ramifications of a Metaphor*. Chicago: University of Chicago Press, 2016.

Ward, Nathaniel Bagshaw. *On the Growth of Plants in Closely Glazed Cases*. London: John Van Voorst, 1842.

Warner, Michael. *The Letters of the Republic: Publication and the Public Sphere in Eighteenth-Century America*. Cambridge, MA: Harvard University Press, 1990.

Waters, James. "Letter from James Waters to Sir Joseph Hooker, October 6, 1851." In *Director's Correspondence at Kew Gardens*, vol. 31. Kew Gardens, London.

W.D. "Transmutation of Plants." *Farmer's Cabinet* 15 (March 3, 1838).

Weheliye, Alexander. *Habeas Viscus: Racializing Assemblages, Biopolitics, and Black Feminist Theories of the Human*. Durham, NC: Duke University Press, 2014.

Wexler, Laura. *Tender Violence: Domestic Visions in an Age of U.S. Imperialism*. Chapel Hill: University of North Carolina Press, 2000.

Whitman, Walt. "This Compost." In *The Complete Poems of Walt Whitman*. New York: Penguin, 2005.

Whyte, Kyle. "Indigenous Climate Change Studies: Indigenizing Futures, Decolonizing the Anthropocene." *English Language Notes* 55, nos. 1–2 (2017): 153–162.

———. "Is It Colonial Déjà Vu? Indigenous Peoples and Climate Injustice." In *Humanities for the Environment: Integrating Knowledges, Forging New Constellations of Practice*, edited by Joni Adamson, Michael Davis, and Hsinya Huang, 88–104. New York: Routledge, 2016.

Wineapple, Brenda. *Hawthorne: A Life*. New York: Random House, 2004.

Wohlleben, Peter. *The Hidden Life of Trees: What They Feel, How They Communicate*. New York: Greystone, 2015.

Wood, Harriet. "The Burdock and the Violet." *Home Journal*, February 2, 1856.

Woodlin, Johnson. "Communication." *Christian Recorder*, January 26, 1867.

Woods, Rebecca. *The Herds Shot Round the World: Native Breeds and British Empire, 1800–1900*. Chapel Hill: University of North Carolina Press, 2017.

Woodson, Lewis. "Gardening." *Frederick Douglass' Paper*, September 23, 1853.

Wulf, Andrea. *The Founding Gardeners: The Revolutionary Generation, Nature, and the Shaping of the American Nation*. New York: Vintage, 2012.

———. *The Invention of Nature: Alexander von Humboldt's New World*. New York: Vintage, 2016.

Wynter, Sylvia. "Unsettling the Coloniality of Being/Power/Truth/Freedom: Towards the Human, After Man, Its Overrepresentation—An Argument." *CR: The New Centennial Review* 3, no. 3 (Fall 2003): 257–337.

Yao, Christine "Xine." *Disaffected: The Cultural Politics of Unfeeling in Nineteenth-Century America*. Durham, NC: Duke University Press, 2021.

Youmans. "Chemistry Applied to the Mechanic and Farmer." *Southern Planter*, July 1852, 206.

Youth's Companion. "The Sleep of Plants." April 3, 1873.

———. "Variety: Poisonous Plant." 12, no. 2 (May 25, 1838).

Yusoff, Kathryn. *A Billion Black Anthropocenes or None*. Minneapolis: University of Minnesota Press, 2019.

INDEX

Page numbers in *italics* indicate Figures.

ABOUT THE AUTHOR

MARY KUHN is Assistant Professor of Environmental Humanities at the University of Virginia.

Printed in the United States
by Baker & Taylor Publisher Services